Failure and Damage Analysis of Advanced Materials

Failure and Damage Analysis of Advanced Materials

Editor

Ambrish Kumar

scitus
academics

Failure and Damage Analysis of Advanced Materials

Edited by **Ambrish Kumar**

Printed in 2017

ISBN: 978-1-68117-216-3

Library of Congress Control Number: 2015936577

© 2016 by
SCITUS Academics LLC,
616, Corporate Way, Suite 2, 4766,
Valley Cottage, NY 10989

www.scitusacademics.com

Contents

Preface

Failure and damage analysis with discussed on advanced materials such as composites, laminates, sandwiches and foams, and also new metallic materials. Starting from some mathematical foundations (limit surfaces, symmetry considerations, invariants) new experimental results and their analysis are shown. Finally, new concepts for failure prediction and analysis will be introduced and discussed as well as new methods of failure and damage prediction for advanced metallic and non-metallic materials. Based on experimental results the traditional methods will be revised. Failure as a limit state of the material behavior is well known from engineering practice. Different types of failure can be identified: transition from the elastic to plastic state, loss of stiffness, loss of fracture resistance at different scale levels, ultimate strength, and fatigue. In addition, failure can be accompanied by various types of damage.

Editor

Durability Modeling of Environmental Barrier Coating (EBC) Using Finite Element Based Progressive Failure Analysis

Ali Abdul-Aziz[1], Frank Abdi[2], Ramakrishna T. Bhatt[1], and Joseph E. Grady[1]

[1]NASA Glenn Research Center, Cleveland, OH 44135, USA
[2]AlphaSTAR Corporation, Long Beach, CA 90804, USA

ABSTRACT

The necessity for a protecting guard for the popular ceramic matrix composites (CMCs) is getting a lot of attention from engine manufacturers and aerospace companies. The CMC has a weight advantage over standard metallic materials and more performance benefits. However, these materials undergo degradation that typically includes coating interface oxidation as opposed to moisture induced matrix which is generally seen at a higher temperature. Additionally, other factors such as residual stresses, coating process related flaws, and casting conditions may influence the degradation of their mechanical properties. These durability considerations are being addressed by introducing highly specialized form of environmental barrier coating (EBC) that is being developed and explored in particular for high temperature applications greater than 1100°C. As a result, a novel computational simulation approach is presented to predict life for EBC/CMC specimen using the finite element method augmented with progressive failure analysis (PFA) that included durability, damage tracking, and material degradation model. The life assessment is carried out using both micromechanics and macromechanics properties. The macromechanics properties yielded a more conservative life for the CMC specimen as compared to that obtained from the micromechanics with fiber and matrix properties as input.

INTRODUCTION

Durability and damage related issues concerning fiber reinforced ceramic matrix composites (FRCMC), specifically SiC fiber reinforced SiC matrix composites (SiC/SiC), are of significance for low maintenance, dependability, and cost efficiency. Typically, most of damage and failure are caused by environmental conditions. These conditions are confined to moisture, thermal-mechanical load, creep, and fatigue. Lab, burner rig, and field tests have been performed to capture the service environment, induced damage, and resulting strength/stiffness reduction for several classes of CMCs being considered as components in aeroengines [1]. The CMCs are lightweight materials and operate at higher temperature than metals of at least 200°C. In dry air conditions, these materials form a protective layer on the surface called silica which makes them stable at a temperature up to 1300°C for long-term applications. However, in combustion environment containing moisture, the silica layer disintegrates causing surface recession [2]. Therefore, in order for these CMC materials to be useful in aeroengine applications, their surface must be protected. Such protection is being considered by applying environmental barrier coating (EBC) that has a range of operating temperature between 1200 and 1500°C depending on the composition [2–6].

There are three classes of EBC currently being evaluated for SiC/SiC turbine components. They are barium aluminum strontium (BSAS), rare earth di- and monosilicates (REMS and REDS), and hafnia/zirconia based systems [7–9]. The rare earth series include elements from lanthanum to lutetium. In general, an EBC system consists of two or more layers of coating, in which each layer serves a specific purpose. The total thickness of the EBC applied depends on the components and the coating can be applied by different processing methods depending on the intended microstructure and durability. Static components such as combustor liners, turbine vanes, and shrouds are subjected to thermal and gas pressure loads only. As a result, these components can accommodate coating thickness as much as 525 µm. On the other hand, rotating components such as blades are subjected to a combination of thermal and mechanical loads. To reduce overall weight of the rotating component, the thickness of EBC is limited to ~125 µm [10].

The coating system can be applied via variety of application systems. Among the most common ones are techniques such as air plasma spray (APS), physical vapor deposition (PVD), and slurry depending on the components, manufacturing cost, and intended durability. These systems in general have different material properties than the substrate since the sublayers of the coating are applied at different temperatures and as a result residual stresses develop. Depending on the magnitude and nature of these stresses, damage can occur in the coating after deposition and after exposing the coated substrate to turbine operating conditions. The damage has to be minimized or controlled; otherwise, the coating will spall which will reduce and limit the life of the component [7–9].

Therefore, damage drivers such residual stresses are of concern to EBC development, durability, and application. These stresses can be determined or measured by nondestructive techniques such as X-ray diffraction, Raman spectroscopy, and neutron diffraction. However, because of the complex crystal structure of some of the EBC compositions, it is cumbersome to use these techniques for these measurements. A means to tackling these factors is to control the constituent properties and thickness of the coating and develop physics based models that enable prediction of the durability and service life of the EBC under typical environmental conditions such as moisture, creep, fatigue, and crack propagation at the coating-CMC interface. PFA is used to determine the residual stresses in the specimen and to evaluate their role in damage initiation and propagation.

This paper is an extension of a prior work [1] where the focus of the research was based on examining an analytical methodology to model the durability of the EBC using a multiscale progressive failure analysis [14,15] approach. Prior work detected damage initiation events using lamina fatigue properties for the CMC as input to the PFA analysis. However, in CMC composites damage initiates at the microscale level of the material. The use of fiber and matrix constituent properties enables the evaluation of damage events at their inception source. With macromechanics, the lamina properties are degraded at the onset of damage. But, with micromechanics, the properties of the constituent that is damaged are degraded, while the other constituent retains its properties.

The analysis used an updated material model for the EBC as compared to the one used in the prior work. Also, it used strength-time exposure degradation model from literature with improved accuracy for the SiC/SiC CMC material. The life prediction was performed once using reverse engineered micromechanics properties as input and once more using macromechanics properties as input. The lamina properties consisted of stiffness, strength, and fatigue properties. Similarly, derived fiber matrix properties consisted of stiffness, strength, and fatigue properties for each constituent. Strength based failure criteria based on maximum stress were employed to determine material damage. Stiffness of damaged elements is reduced once a specific failure criterion is invoked. Damaged elements are not removed from the finite element model. Future work can use fracture mechanics principle using the damage path predicted by PFA to assess fracture growth in the EBC coated CMC specimen. Results from the analytical effort are discussed next.

DESCRIPTION OF ANALYTICAL APPROACH

The analyses utilized the finite element method to model the combined EBC/CMC bar specimen sample which included the three layers of EBC and the coated substrate or the CMC part. Finite element model was developed for estimating the stress response based on the known processing conditions of the coating, the specimen geometry of the coated substrate, and the thermomechanical properties of the coated layers and the substrate.

Coating was applied via the plasma spray technique on SiC/SiC composite substrates [9]. It is assumed that the substrate is maintained at 1300°C during deposition of the plasma spray coating and then cooled to room temperature ~25°C. Also, thermomechanical properties of standalone individual layers of the EBC system required for the model were obtained from [13].

The analytical calculations covered modeling the beam specimen with defined EBC layers on top of a SiC/SiC substrate. Thermomechanical properties, including thermal expansion coefficient, stiffness, Poisson's ratio, and strength for all four materials constituting the EBC/CMC specimen, are used as input to the durability and life prediction analysis; see Tables 1 and 2. The SiC/SiC CMC material lamina properties of the fabric were obtained from [12]. Plastic deformation and microcracking that may occur in the plasma-sprayed coating were not considered in the model. The specimen dimensions were 2 by 3 by 45 mm in addition to the EBC thickness on the top, Figure 1. The thermal boundary conditions associated with the coating application methodology were all incorporated into the thermal model.

Table 1: Physical, thermal, and mechanical properties of uncoated SiC/SiC substrate at 21°C [12]

SiC/SiC property	Value	SiC/SiC property	Value
E11 (MPa)	$2.85E + 05$	$S11T$ (MPa)	$3.21E + 02$
E22 (MPa)	$2.85E + 05$	$S11C$ (MPa)	$3.21E + 02$
E33 (MPa)	$1.57E + 05$	$S22T$ (MPa)	$3.21E + 02$
G12 (MPa)	$1.13E + 05$	$S22C$ (MPa)	$3.21E + 02$
G23 (MPa)	$9.90E + 04$	$S33T$ (MPa)	$4.00E + 01$
G13 (MPa)	$9.90E + 04$	$S33C$ (MPa)	$1.00E + 02$
NU12	$1.30E - 01$	$S12S$ (MPa)	$2.10E + 02$
NU23	$1.70E - 01$	$S23S$ (MPa)	$1.05E + 02$
NU13	$1.70E - 01$	$S13S$ (MPa)	$1.05E + 02$
ALPHA11 (1/°C)	$2.71E - 06$	ALPHA22 = 33 (1/°C)	$2.71E - 06$

Table 2: Material properties of top and intermediate coats and bond [13]

(a)

Bond coat properties	Value	Units
Young's modulus	97	(GPa)
Poisson's ratio	0.21	(—)
CTE	4.50 – 06	(1/°C)
Tension strength	40	(MPa)
Compression strength	40	(MPa)
Shear strength	40	(MPa)

(b)

Intermediate coat properties	Value	Units
Young's modulus	37.4	(GPa)
Poisson's ratio	0.179	(—)
CTE	5.70 – 06	(1/°C)
Tension strength	28	(MPa)
Compression strength	28	(MPa)
Shear strength	28	(MPa)

(c)

Top coat properties	Value	Units
Young's modulus	32	(GPa)
Poisson's ratio	0.19	(—)
CTE	5.60 – 06	(1/°C)
Tension strength	28	(MPa)
Compression strength	28	(MPa)
Shear strength	28	(MPa)

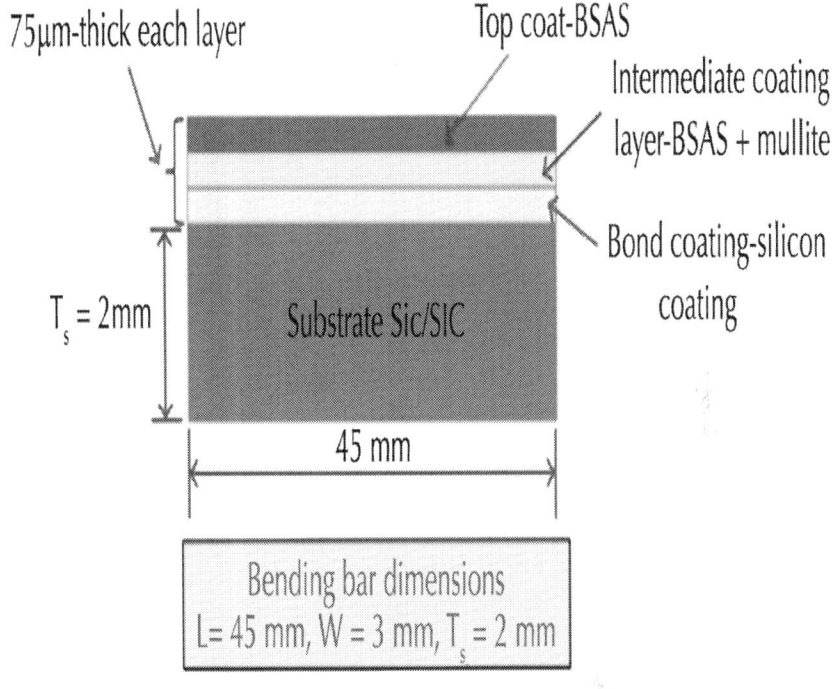

Figure 1: Two-dimensional section of the beam bar specimen showing dimension.

Multiscale Progressive Failure Analysis (PFA)

Micromechanics and macromechanics composite analysis are integrated with finite element analysis and damage and fracture tracking to perform progressive failure analysis, Figure 2. The capability is integrated in the GENOA [14, 15] software. Traditionally, failure is assessed at the macroscale using lamina or laminate properties. The software enables assessment of failure and damage in composites at a lower scale, that is, the fiber, matrix, and interface level. The methodology augments finite element analysis (FEA), with a full-hierarchical modeling capability that goes down to the microscale of subdivided unit cells composed of fiber bundles and their surrounding matrix [16]. The life prediction strategy uses a PFA-FEA based approach shown in Figure 2 [14, 15].

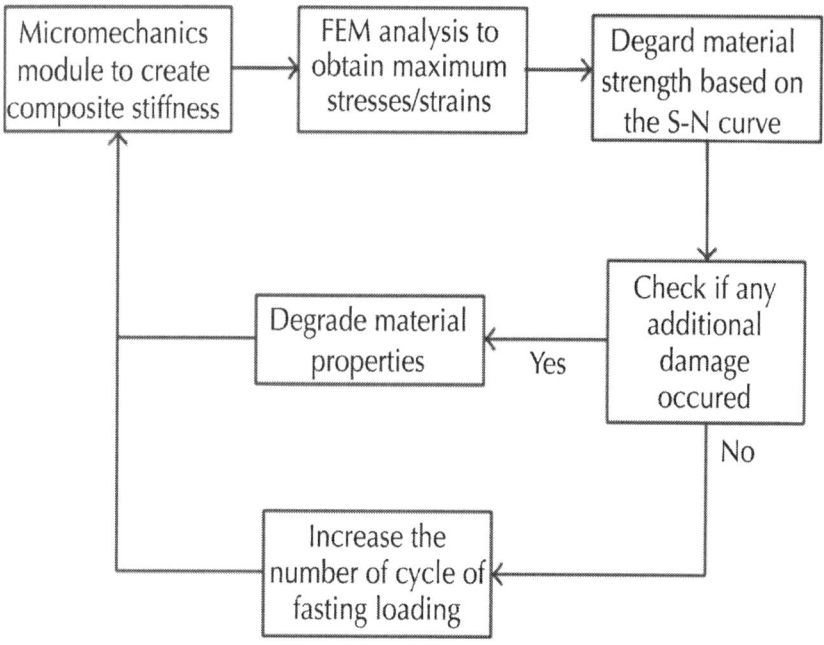

Figure 2: General flow of the progressive failure analysis methodology for life prediction [1].

Damage is tracked at the micro- or macroscale levels leading to local material degradation and recalculation of stiffness. This is done by evaluating series of physics based failure criteria (shown in Table 3) at increased load increments or fatigue cycles. In addition to degradation, stress damage, strength-cycle, or strength-time curves are used as input to the analysis at each fatigue cycle block to degrade the strength. Damage is accumulated as life cycles are increased until the ultimate life of the structure is reached.

Table 3: Failure criteria used in life prediction of composite specimens

Mode number	Fiber failure criteria	Event description
1	Longitudinal tensile (11T)	Failure of ply controlled by fiber tensile strength and fiber volume ratio

2	Longitudinal compressive (11C)	(1) Fiber/matrix delamination under compression loading (2) Fiber microbuckling (3) Fiber crushing
3	Transverse tensile Transverse compressive Normal tensile (22T)	Matrix cracking under tensile loading, event controlled by matrix tensile strength, matrix modulus, and fiber volume ratio Matrix cracking under compressive loading, event controlled by matrix compressive strength, matrix modulus, and fiber volume ratio Plies are separating due to normal tension
4	Normal compressive (22C)	Due to very high surface pressure that is crushing of laminate
5	In-plane shear (33C)	Failure in-plane shear relative to laminate
6	Transverse normal shear (11)	Shear failure acting on transverse cross-oriented in a normal direction of the ply
7	Longitudinal normal shear (12S) Normal tensile (13S) Relative rotation criterion (RROT)	Shear failure on longitudinal cross section that is oriented in a normal direction of ply Combined stress failure criteria used for isotropic materials Considers failure if the adjacent plies rotate excessively with one another
8	Transverse normal shear (23S)	Considers invariant through-the-thickness
9	Linear elastic fracture	Virtual crack closure technique (VCCT), discrete cohesive zone model (DCZM)

The life prediction analysis uses PFA to determine how many cycles of temperature ramping the specimen can sustain before the SiC/SiC and the coating materials are damaged; see Figure 2. Ideally, strength-time curves would be required as input to the analysis for all the materials constituting the EBC/CMC specimen. Such data are typically obtained from physical testing or from literature. The analysis assumed that the EBC coating materials do not degrade as function of exposure time to temperature. Only the SiC/SiC CMC is degraded using test data obtained from literature for strength degradation as function of time [11].

Life Prediction with PFA

The mathematical approach used in applying the PFA includes the integration of composite mechanics and damage/fracture mechanics with finite element analysis. The damage mechanics account for matrix cracking under transverse, compressive, and shear loading. The ply fracture mechanisms include fiber failure under tension, compression (crushing, microbuckling, and debonding), and delamination. This is invoked via the GENOA code [16] by allowing a sequence of analytical steps that includes (1) the use of a finite element stress solver, (2) user selection of 2D or 3D architectural details (through-the-thickness fibers, resin rich interphase layer between weave plies, fiber volume ratio, void shape, size and location, cure condition, etc.), (3) assigning static (thermomechanical) or spectrum loading, (4) automatic update of the finite element model prior to executing FEA stress solver for accurate lamina and laminate properties, and (5) degradation of material properties at increased loading (including number of cycles) based on detected damage. Additional details can be found in [14].

All stages of damage evolution within the composite structure are identified. They are damage initiation, damage propagation, fracture initiation, and fracture propagation. The damage events include matrix cracking, delamination, and fiber failure. Displacements, stresses, and strains derived from the structural scale FEA solution at a node or element of the finite element model are passed to the laminate and lamina scales using laminate theory. Stresses and strains at the microscale are derived from the lamina scale using microstress theory. The analytical capability offers microscale modeling and damage assessment capability for composite materials such as ceramic, metal, and polymer matrix composites. For the failure criteria shown in Table 3, the code automatically distinguishes between the criteria that are applicable to laminated composites versus those that are applicable to isotropic material. For example, for longitudinal compression, the code evaluates three failure potentials under longitudinal compression. They are fiber and matrix delamination, fiber microbuckling, and fiber crushing. For isotropic materials, the compression stress or strain is compared to the allowable to determine whether or not failure had occurred. In addition to maximum stress failure criteria, the code evaluates failure due to maximum strain and interactive stress criteria. More details can be obtained from [14, 15].

PFA stress based evaluation is accurate up to fracture initiation [14, 15]. Due to stress singularity often experienced in finite element analysis, fracture mechanics based approach is used to grow the crack. In linear fracture mechanics approach, it is required to have fracture toughness for static crack growth and da/dn versus ΔK for fatigue crack growth; da/dn is the change of crack length with loading cycles, while ΔK is the stress intensity factor change. Since such data are not available for the analysis, the focus in this paper is to identify cycles that caused damage to initiate and propagate to the substrate SiC/ SiC material. It should be noted that the PFA strength based approach has key advantages as compared to fracture mechanics methods. For example, in terms of advantages, PFA does not require prior knowledge of crack path. Crack growth will be the subject of future work once the data required for the prediction becomes available.

Additionally, for the analyses used in this paper, linear elastic fracture mechanics approach (item no. 9 in Table3) is not used. Fracture mechanics application would require knowledge of fracture path, toughness, and fracture energy. This type of analysis will be considered in the future.

SPECIMEN GEOMETRY AND FE MODEL

The finite element analyses to generate the thermal profile were conducted using the commercial finite element code Abaqus [17]. The finite element model dimensions and sections are shown in Figure 3.

The thermal profile was predicted under transient loading conditions as noted in Figure 4. The thermal cycle assumes that the bar specimen is initially at 21°C and, within 15 minutes, it heats up to 1300°C and remains constant for a duration of 45 minutes until shutdown, where it cools off back to 21°C. One complete cycle constitutes exposing the specimen to these thermal conditions for a total time of one hour. Material properties of both the coating and the substrate were input into the model under linear isotropic condition for the coating systems and linear orthotropic condition for the SiC/SiC substrate. Temperature dependency of all the materials was accounted for in the analyses. The mesh included a 3D model of the bar specimen with high density mesh along the substrate and the coatings interfaces. Eight-node brick element was employed.

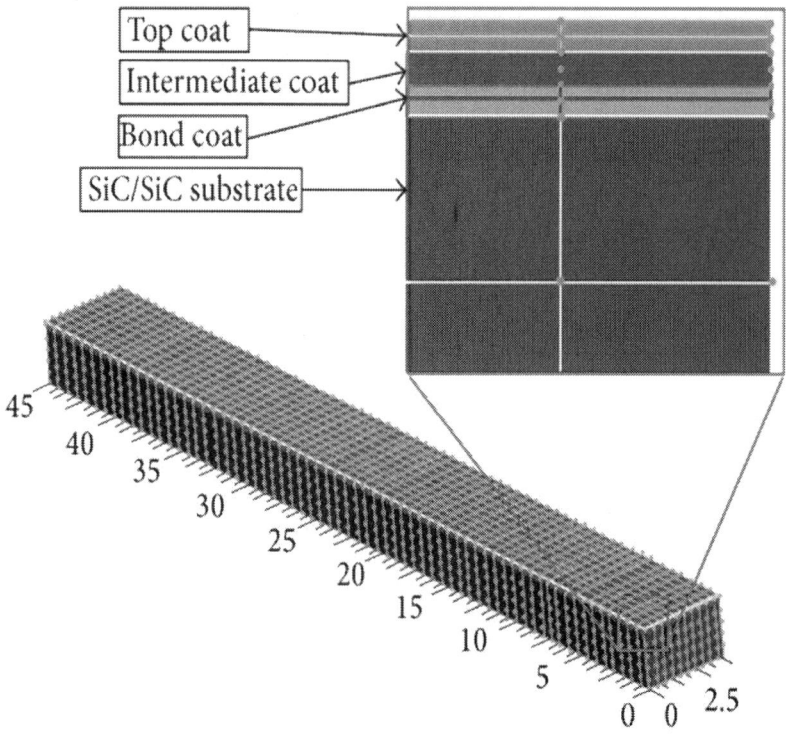

Figure 3: Representative finite element model of the thermal barrier coating. Work plane rulers shown are in units of mm.

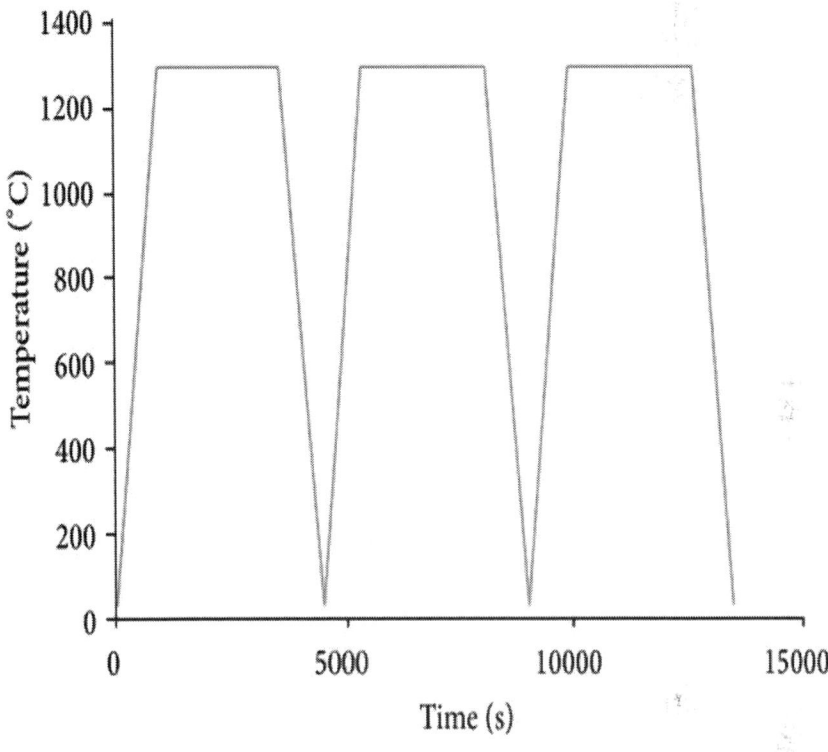

Figure 4: Representative thermal loading profile applied in the analyses.

For the durability solution, GENOA-PFA augmented the FEA solver (Abaqus) for life prediction to determine life cycles that caused damage to initiate and propagate. PFA used Abaqus iteratively at increased number of cycles to evaluate damage after each FEA run. Damage stabilization is attained to ensure material and structural equilibrium before the number of cycles is increased again. The process is repeated until ultimate life is obtained. It was assumed that the top coat BSAS, intermediate coating layer (BSAS+Mullite), and the bond silicon coating do not degrade as a function of time and temperature and the results presented pertained only to determining the number of cycles that it would take for damage to initiate and for damage to propagate. Furthermore, data obtained from literature [11] shows that some degradation of the SiC/SiC substrate at elevated temperature after exposure time is expected.

ANALYTICAL DESCRIPTION

To evaluate the effects of thermal fatigue on the EBC SiC/SiC using micromechanics based approach, in situ (effective) fiber and matrix properties for the SiC/SiC system were generated. To generate the effective SiC/SiC constituent properties, a [0,90]s laminate with 50% fiber volume content was modeled using materials characterization and qualifications (MCQ) composite software [18] and an iterative process (Figure 5) was implemented. The code used an optimization algorithm to derive a unique set of fiber/matrix properties (mainly stiffness and strength) that are capable of reproducing test data of the lamina or laminate. The process iterates until values of the predicted lamina or laminate in-plane and out-of-plane properties are in good agreement with data from test. The fiber and matrix properties for the SiC/SiC derived through the elaborate reverse engineering process shown in Tables 4(a) and 4(b) are used as input to PFA to determine the life cycle that would cause damage to initiate and propagate in the CMC specimen. This makes it possible to run the comparative assessment of the behavior of the CMC using both microscale (fiber and matrix) and macroscale (lamina level). PFA treats the top, intermediate, and bond materials as isotropic during the life prediction analysis.

Table 4: (a) Effective fiber and matrix properties (b) Effective ply correlating to effective fiber/matrix properties

(a)

Effective fiber and matrix properties (FVR = 0.5[0,90] s)					
SiC fiber			SiC matrix		
Property	Units	Value	Property	Units	Value
E11	MPa	380000	E	MPa	380000
E22	MPa	156000	NU		0.19
G11	MPa	100000	ST	MPa	250
G23	MPa	70000	SC	MPa	250
NU12		0.19	SS	MPa	190
NU23		0.17	ALPHA	1/degC	2.71 − 06

S11T	MPa	600			
S11C	MPa	600			
ALPHA11	1/degC	2.71 − 06			
ALPHA22	1/degC	2.71 − 06			

(b)

Properties of[0,90] s using effective fiber/matrix properties				
Property	Units	Target values	Calibrated	Discrepancy (%)
E11	MPa	284900	285000	0.0
E22	MPa	284900	285000	0.0
E33	MPa	190600	157000	17.6
G12	MPa	112300	113000	0.6
G23	MPa	104800	99000	5.5
G13	MPa	104800	99000	5.5
NU12	MPa	0.126	0.13	3.2
NU23	MPa	0.2132	0.17	20.3
NU13	MPa	0.2132	0.17	20.3
S11T	MPa	318.5	321	0.8
S11C	MPa	318.5	321	0.8
S22T	MPa	318.5	321	0.8
S22C	MPa	318.5	321	0.8
E12S	MPa	213.5	210	1.6

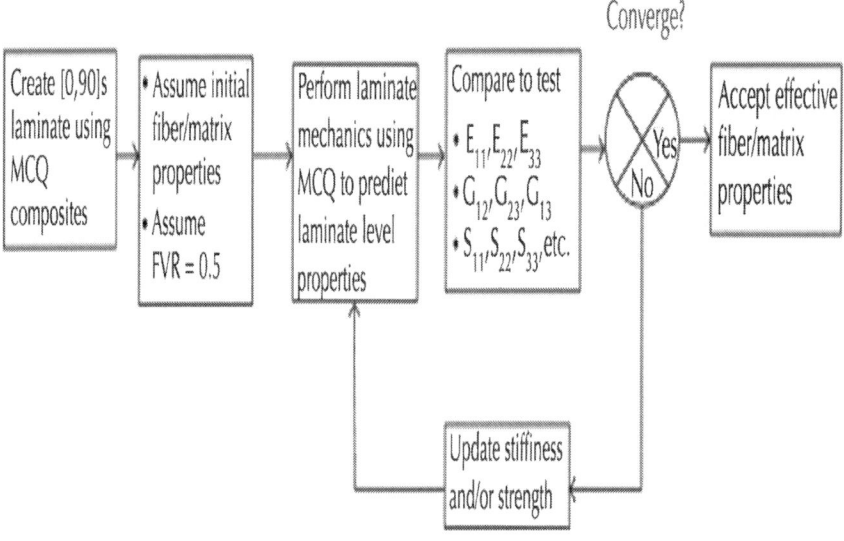

Figure 5: Iterative process used to determine effective constituent properties of SiC/SiC laminate.

The fiber and matrix properties obtained from reverse engineering are listed in Table 4(a). The fiber and matrix properties are then used as input to MCQ [18] to evaluate the mechanical properties of the CMC (0/90)s laminate. As indicated in Table 4(b), the MCQ predictions starting from microscale properties yielded an accurate representation of the laminate properties from test [12]. The out-of-plane predictions for the laminate could improve as the properties were taken from a plain weave system where the effects of the fiber weave on the out-of-plane strength and stiffness was significant. If more details were available on the architecture, the out-of-plane predictions with MCQ would improve as the architecture details would be included in material characterization.

SiC/SiC Stress-Cycle Curve as Function of Exposure Time

The stress cycle (S-N) criterion adopted for these analyses utilized a set of S-N curve for each of the laminate, fiber, and matrix as shown in Figure 6. The first set used the typical or the standard S-N curve that represented the degradation of the SiC/SiC composite under thermal loading conditions [11]. This S-N curve started to degrade after 27.78 hours from 318 MPa to 159 MPa (50% of the original ultimate tensile strength at t=0) at 1000 hr. The logarithmic degradation continued until failure. Since S-N curve was not available for the constituent materials, it was assumed that the same degradation trends occurred for the fiber and matrix level. To achieve these trends, the laminate level trend was scaled accordingly using the fiber and matrix ultimate tensile strength at t=0.

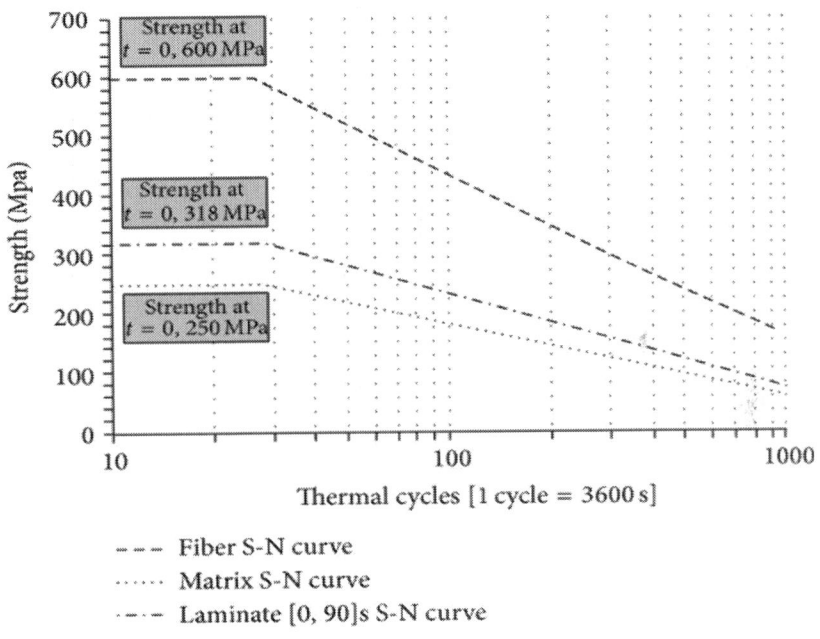

Figure 6: Strength as a function of thermal cycles for laminate, fiber, and matrix of SiC-SiC material [11].

Characteristics and thickness dimensions of the coating system used are shown in Table 5 and, as noted, the 3 layers of coating had the same magnitude of thickness which is 75 μm and the substrate had a 2 mm thickness. Photographs of the top surface and cross section of a typical APS trilayered environmental barrier coated SiC/SiC composite are shown in Figure 7. The sublayers of the coating are inhomogeneous and contain microcracks and significant levels of nonuniform pores.

Table 5: Coating systems and thicknesses considered

Coating system	Coating thickness
Top coat-BSAS	75 µm
Intermediate coat Mullite + barium strontium aluminum silicate (BSAS)	75µm
Bond coat-silicon	75µm
Sic-SiC substrate	2 mm

(a)

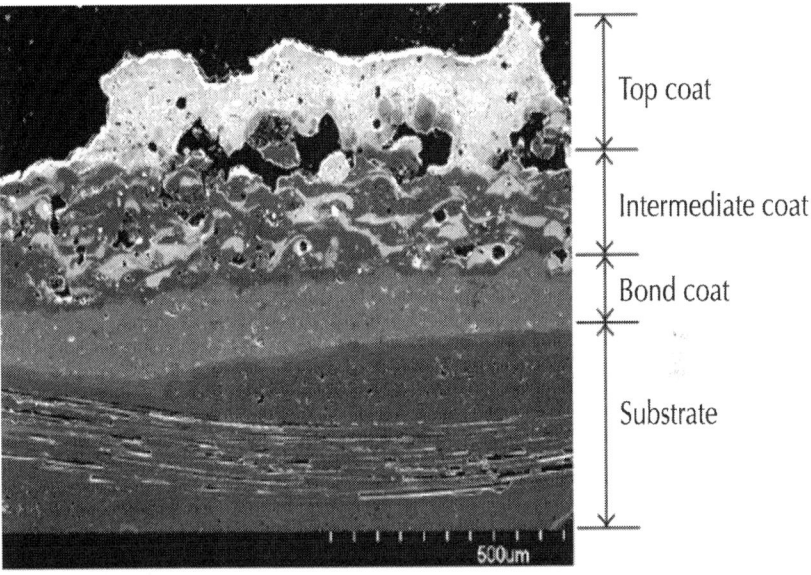

Top coat

Intermediate coat

Bond coat

Substrate

(b)

Figure 7: Typical microstructure of plasma sprayed EBC on SiC/SiC composites: (a) top view (optical micrograph) and (b) cross-sectional view (scanning electron micrograph).

RESULTS AND DISCUSSION

The durability analysis performed indicated that the material damage initiated in the top EBC coat and then propagated down to the intermediate layer then to the bond. This took place during the first one hour of thermal loading (in one cycle). For each cycle, the PFA analysis assumed that the specimen reached the 1300°C temperature, which means that a gradient of 1279°C is applied instantly to simulate each cycle.

As mentioned earlier in the paper, the maximum stress criteria listed in Table 3 are used in the durability evaluation performed by the PFA. When an element stress exceeds the allowable value, the stiffness of the element was degraded accordingly in the direction where the damage occurred. No elements were removed when damaged, which meant that the mesh size remained unchanged throughout the analysis. The analysis was repeated twice, once with macrolevel properties for the CMC material and once more with microlevel properties using the reverse engineered properties of the fiber and matrix.

Figure 8 shows the damaged elements in red color in the three EBC layers due to tension stress. As noted, all elements in the three EBC layers are damaged at the end of cycle 1, that is, after 3600 seconds of exposure to 1300°C. The damage started in the top layer and propagated down to the bond. The SiC/SiC substrate was undamaged until the cycle 1070 was reached for macrotype input and until the cycle 1090 for micro input.

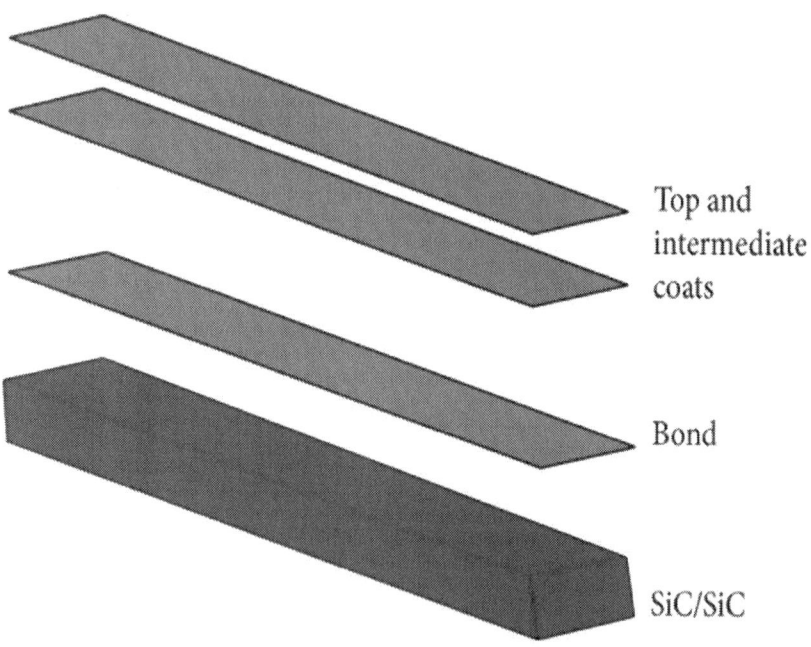

Top and intermediate coats

Bond

SiC/SiC

Figure 8: Damage in the EBC layers due to tension stress (red color indicates damage; blue color represents undamaged elements).

The damage volume is computed to keep track of total number of elements that are damaged during a given loading cycle. It provides useful inspection criterion of critical parts. For example, a sudden increase in damage volume indicates the onset of major damage event in the structure. Since only degradation in the SiC/SiC substrate was considered, no additional damage was detected until cycle 1070 when the SiC/SiC substrate's tensile failure criterion was detected. A summary of the life cycles for the EBC-SiC/SiC system is shown in Table 6. A macrobased simulation seemed to offer more conservative life cycle compared with the microbased approach showing greater life by approximately 20 hours. This is assuming the same in-plane ply properties and when using effective fiber/matrix properties rather than macrobased laminate properties.

Table 6: Summary of cycles to damage and associated damage modes for each EBC and substrate layer

Layer	Cycle of damage initiation	Damage mode
Top layer	1	Tension
Intermediate layer	1	Tension and shear
Bond layer	1	Tension and shear
SiC/SiC	Macro: 1070	Tension
	Micro: 1090	

The results indicated that macromechanics or ply mechanics approach is more conservative when it comes to assessing damage initiation in the substrate as compared to microscale simulation. This was expected as the postdamage degradation was more severe at the macrolevel as compared to the one at the microscale. With micromechanics, if one constituent is damaged, the other constituent retains the stiffness. In the case of macromechanics approach, the whole lamina stiffness in the direction of damage is reduced to a small value.

Figure 9 shows the von Mises residual stresses in each material obtained during cool down from 1300°C to room temperature. The residual stresses are a good indicator of where damage is likely to start. It supports the findings presented in Figure 9, whereby the top, intermediate, and bond layers were damaged first before propagating after several hundred cycles to the CMC substrate. The PFA analysis is accounting for the residuals' stresses during the life prediction as the residual stresses are translated into damage indices when damage is introduced. The damage indices are then stored for use as input in the subsequent cycle analysis.

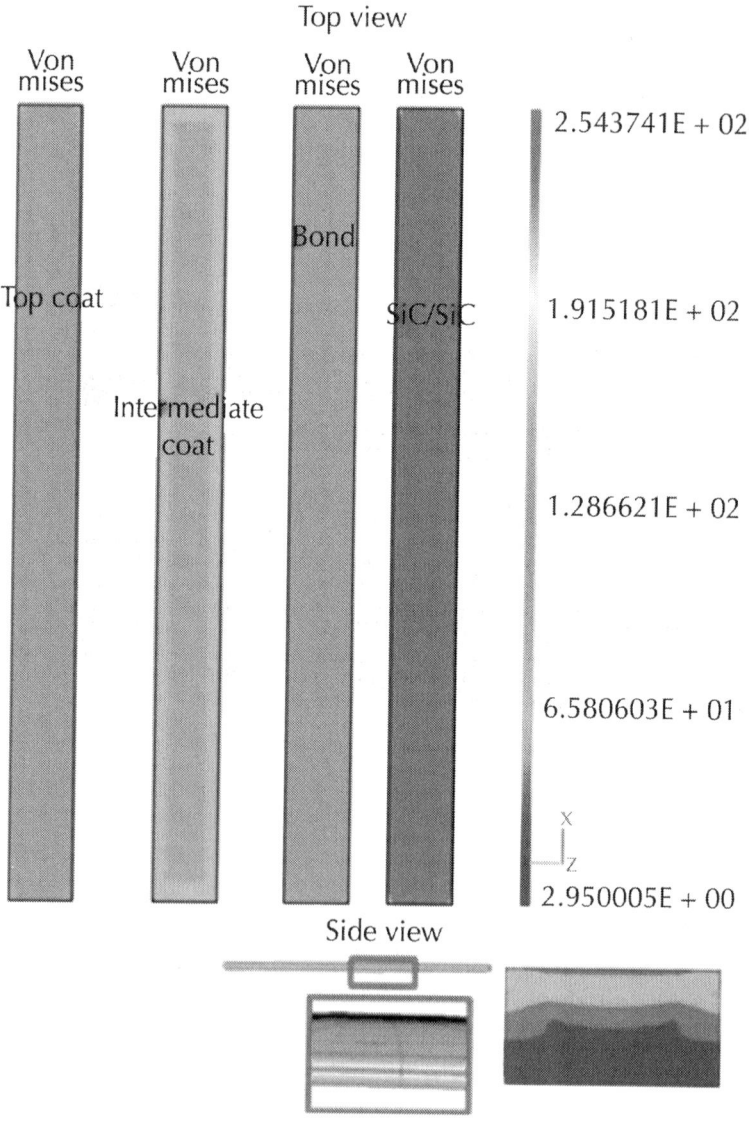

Figure 9: Von Mises residual stresses in MPa as a result of cool down from 1300°C to room temperature after the first cycle of exposure to elevated temperature (top and side views).

It should be further noted that the von Mises stresses calculated during cool down show that the bond material experienced the highest stresses. However, this does not mean that the bond is failing more than the top or intermediate coats. Failure is driven by the allowable material stress or strain. In the case of the EBC specimen, the maximum stress criteria were used to guide the assessment of the damage evolution. Comparing stress to strength for the top three materials, the residual stresses do indicate that material damage is experienced by all three materials, whereby the top coat experienced the most damage because of the ratio of stress to strength. Future analysis will include material nonlinearity as well as coupled structural thermal analysis to determine the effect of nonuniform heating on the specimen's life prediction.

CONCLUSIONS AND FUTURE WORK

A novel computational simulation approach is presented to predict life for EBC/CMC specimen using the finite element method augmented with progressive failure analysis (PFA) that included durability, damage tracking, and material degradation model. The following conclusions and recommendations can be drawn from the work presented in the paper.

- Damage initiates predominately in the top coat due to tensile strength failure and in the intermediate/bond coat due to delamination.

- Damage propagates into the SiC/SiC substrate due to tensile failure, eventually redistributing the stress into the EBC causing further damage propagation.

- Multiscale progressive failure analysis allowed a systematic prediction of the life cycles for damage initiation and propagation in EBC SiC/SiC specimens. The technical approach applied combined composite mechanics and damage tracking and fracture.

- Use of micromechanics properties as input to PFA resulted in life prediction that is approximately 20 hours greater as compared to that obtained from the use of lamina properties indicating that macromechanics is more conservative than micromechanics.

- Accurate life prediction requires strength-time exposure behavior for all the materials used in the specimen. This will allow reliable assessment of any structural component made of the same materials.

- Defects such as flaws and initial cracks in coating will add more accuracy to the life prediction analysis and it all must be accounted for in any future work.

- Material characterization can help optimization of the laminate thickness which in return can increase life and delay damage.

- Material architecture should be considered in the material characterization to yield an accurate reverse engineering of constituent properties.

- Future work should include nonlinear material behavior in the analysis and simulations performed. This will require data from ATSM tests at different temperatures.

REFERENCES

1. A. Abdul-Aziz, G. Abumeri, W. Troha, R. T. Bhatt, J. E. Grady, and D. Zhu, "Environmental barrier coating (EBC) durability modeling using a progressive failure analysis approach," in Smart Structures and Materials & Nondestructive Evaluation and Health Monitoring, Behavior and Mechanics of Multifunctional and Composite Materials, vol. 8346 of Proceedings of SPIE, San Diego, Calif, USA, March 2012.

2. K. N. Lee, D. S. Fox, R. C. Robinson, and N. P. Bansal, "Environmental barrier coatings for silicon-based ceramics," in High Temperature Ceramic Matrix Composites, W. Krenkel, R. Naslain, and H. Schneider, Eds., pp. 224–229, Wiley-Vch, Weinheim, Germany, 2001.

3. P. J. Jorgensen, M. E. Wadsworth, and I. B. Cutler, "Oxidation of silicon carbide," Journal of the American Ceramic Society, vol. 42, no. 12, pp. 613–616, 1959.

4. J. L. Smialek, R. C. Robinson, E. J. Opila, D. S. Fox, and N. S. Jacobson, "SiC and Si_3N_4 recession due to SiO_2 scale volatility under combustor conditions," Advanced Composite Materials, vol. 8, no. 1, pp. 33–45, 1999.

5. K. L. More, P. F. Tortorelli, and L. R. Walker, "Effects of high water vapor pressures on the oxidation of SiC-based fiber-reinforced composites," Materials Science Forum, vol. 369—372, pp. 385–394, 2001.

6. K. L. More, P. F. Tortorelli, L. R. Walker, N. Miriyala, J. R. Price, and M. Van Roode, "High-temperature stability of SiC-based composites in high-water-vapor-pressure environments," Journal of the American Ceramic Society, vol. 86, no. 8, pp. 1272–1281, 2003.

7. K. N. Lee, "Current status of environmental barrier coatings for Si-based ceramics," Surface and Coatings Technology, vol. 133-134, pp. 1–7, 2000.

8. D. M. Zhu, R. A. Miller, and D. S. Fox, "Thermal and environmental barrier coating development for advanced propulsion engine systems," NASA TM-2008-215040, 2008.

9. D. M. Zhu, N. P. Bansal, and R. A. Miller, "Thermal conductivity and stability of HfO_2-Y_2O_3 and $La_2Zr_2O_7$ evaluated for 1650°C," in Advances in Ceramic Matrix Composites, N. P. Bansal, J. P. Singh, W. M. Kriven, and H. Schnneider, Eds., John Wiley & Sons, Hoboken, NJ, USA.

10. D. Zhu and R. A. Miller, "Thermal conductivity and elastic modulus evolution of thermal barrier coatings under high heat flux conditions," Journal of Thermal Spray Technology, vol. 9, no. 2, pp. 175–180, 2000.

11. S. Ochiai, S. Kimura, H. Tanaka et al., "Degradation of SiC/SiC composite due to exposure at high temperatures in vacuum in comparison with that in air," Composites A: Applied Science and Manufacturing, vol. 35, no. 1, pp. 33–40, 2004.

12. M. van Roode, A. K. Bhattacharya, M. K. Ferber, and F. Abdi, "Creep resistance and water vapor degradation of sic/sic ceramic matrix composite gas turbine hot section components," in ASME Turbo Expo 2010: Power for Land, Sea, and Air, pp. 455–469, June 2010.

13. A. Abdul-Aziz and R. T. Bhatt, "Modeling of thermal residual stress in environmental barrier coated fiber reinforced ceramic matrix composites," Journal of Composite Materials, 2011.

14. M. Garg, G. H. Abumeri, and D. Huang, "Predicting failure design envelop for composite material system using finite element and progressive failure analysis approach," in Proceedings of the 52nd International SAMPE Symposium: Material and Process Innovations: Changing our World (SAMPE '08), May 2008.

15. F. Abdi, Z. Qian, and M. Lee, The Premature Failure of 3D Woven Composites, ACMA Composites, Columbus, Ohio, USA, 2005.

16. "GENOA durability and damage tolerance and life prediction software," AlphaSTAR Corporation, Long Beach, Calif, USA, http://www.ascgenoa.com.

17. "Abaqus commercial finite element code," Providence, RI, USA, 2909—2499.

18. "MCQ software," AlphaSTAR Corp, Long Beach, Calif, USA, 2012.

Chapter 2

FE Analysis of Dynamic Response of Aircraft Windshield against Bird Impact

Uzair Ahmed Dar, Weihong Zhang, and Yingjie Xu

Laboratory of Engineering Simulation and Aerospace Computing, Northwestern Polytechnical University, Xi'an, Shaanxi 710072, China

ABSTRACT

Bird impact poses serious threats to military and civilian aircrafts as they lead to fatal structural damage to critical aircraft components. The exposed aircraft components such as windshields, radomes, leading edges, engine structure, and blades are vulnerable to bird strikes. Windshield is the frontal part of cockpit and more susceptible to bird impact. In the present study, finite element (FE) simulations were performed to assess the dynamic response of windshield against high velocity bird impact. Numerical simulations were performed by developing nonlinear FE model in commercially available explicit FE solver AUTODYN. An elastic-plastic material model coupled with maximum principal strain failure criterion was implemented to model the impact response of windshield. Numerical model was validated with published experimental results and further employed to investigate the influence of various parameters on dynamic behavior of windshield. The parameters include the mass, shape, and velocity of bird, angle of impact, and impact location. On the basis of numerical results, the critical bird velocity and failure locations on windshield were also determined. The results show that these parameters have strong influence on impact response of windshield, and bird velocity and impact angle were amongst the most critical factors to be considered in windshield design.

INTRODUCTION

High velocity bird impact is one of the most significant hazards to both civilian and military aircrafts [1]. When fighter military aircrafts are operated at lower altitude and higher speed, the probability of bird impact increases and proves lethal to the safety of pilot and critical aircraft components. Windshield is the exposed part of aircraft and prone to bird impact. In order to ensure the safety of aircraft, the windshield must be capable of withstanding high velocity impact threats. Design of an optimum impact resistant windshield is a challenge and requires extensive experimental testing. The advanced numerical techniques are being adopted as an effective tool to simulate the bird strike event and provide a substitute to excessive costly experimentations. Moreover, it allows analyzing the most stringent impact conditions that cannot be considered in the experiments and provides detailed insight to the impact process which is difficult to observe during experimental testing.

Several researchers such as Zang et al. [2], Samuelson and Sornas [3], and Boroughs [4] carried out finite element analysis to investigate the impact response of windshield against bird strike. McCarty et al. [5, 6] used Materially and Geometrically Nonlinear Analysis (MAGNA), a nonlinear finite element analysis program for designing of windshield and canopy of military aircrafts. Wang et al. [7–9] simulated the failure of aircraft windshield against bird impact by using a modified nonlinear viscoelastic constitutive model together with Zhu-Wang-Tang (ZWT) damage model for the PMMA-based windshield. The constitutive model and its failure criterion effectively predicted the failure of windshield under a range of impact velocities. The authors also examined different critical factors which affect the impact response of aircraft windshield against bird strike. The results proposed that material model for windshield, boundary conditions, mesh density, surrounding structure of windshield, and bird velocity are the critical factors that must be taken into account for FE analysis of aircraft anti-bird design. Guida et al. [10] numerically examined the influence of geometric parameters of windshield and impact parameters under high speed bird impact by using explicit FE code. The results showed that windshield panel dimensions, thickness, and curvature as well as bird velocity, size, and impact angle considerably affect the impact response of windshield. Liu et al. [11] applied smooth-particle-hydrodynamics (SPH) based approach to model the bird against impact on windshield structure by using explicit FE solver. The maximum displacements of various points on camber line of windshield were compared with experimental results and were found to be in good agreement. The results also indicated that SPH solver gives better comparison than Lagrangian solver. Zhu et al. [12] numerically studied the bird impact response on windshield by employing user-defined material subroutine in explicit FE solver. Numerical results of failure modes of the windshield, deformation, displacement, and strain curves of the measured points on the windshield agreed well with the experimental results. Yang et al. [13] carried out FE and experimental investigation to study the structure strength of windshield subjected to bird impact. Based on numerical scheme, the critical impact velocities on certain locations on windshield are determined and verified through experimental results.

The present work aims to numerically model the impact response of windshield and examine the effect of various factors that contribute to optimize the design of certain windshield. Several critical factors such as mass and shape of bird, impact velocity, angle of impact, and impact location were studied and their influence on the dynamic response of windshield was assessed. The numerical model was developed and implemented in explicit finite element hydro code ANSYS AUTODYN.

NUMERICAL MODELING

Windshield Material Model

The windshield considered in this study is monolithic, uniform cross-section, PMMA-based aviation organic glass. The dynamic properties of this material have been extensively studied in the literature and several material models are available for modeling the impact response of this material [7, 14–17]. In this study, the elastoplastic material model along with maximum principal strain failure criterion was defined to predict damage and failure of windshield. The material model was implemented by incorporating isotropic hardening using Von Mises yield criterion together with rate-dependent Cowper Symonds plasticity law. Cowper Symonds strength model defines the yield strength of isotropic strain hardening of strain rate dependent material [10]. The yield surface can be defined as

$$Y = \left(A + B\varepsilon_{\mathrm{pl}}^{n}\right)\left[1 + \left(\frac{\dot{\varepsilon}^{\mathrm{pl}}}{D}\right)^{1/q}\right],$$

(1)

where Y is yield stress of the material, A is yield stress at zero plastic strain, B is strain hardening coefficient, n is strain hardening exponent, and D and q are strain rate hardening coefficients. Von Mises is simple and convenient criterion to apply as it defines a smooth and continuous yield surface with good approximation at high stresses. At given principal stresses σ_1, σ_2, and σ_3, the yield criterion is defined as

$$\left(\sigma_1 - \sigma_2\right)^2 + \left(\sigma_2 - \sigma_3\right)^2 + \left(\sigma_3 - \sigma_1\right)^2 = 2Y^2.$$

(2)

The maximum principal strain criterion implies that if the maximum tensile principal strain exceeds the prescribed limits, then material will instantaneously fail. Failure is predicted when either of the principal strains ε_1 or ε_2, resulting from the principal stresses σ_1 or σ_2, equals or exceeds the maximum strain corresponding to the yield strength σ_y of the material in uniaxial tension or compression. For yielding in tension the minimum principle strain ε_1 would equal the yield strain in uniaxial tension. If the strains are expressed in terms of stress, then.

$$\varepsilon_1 = \frac{\sigma_1}{E} - \frac{v}{E}\left(\sigma_2 + \sigma_3\right),$$

$$\sigma_1 - v\left(\sigma_2 + \sigma_3\right) \leq \sigma_y,$$

$$\varepsilon_{fail} = \varepsilon_{total} - \frac{\sigma_{total}}{E} \quad \sigma > Y.$$

(3)

Table 1: Material properties of windshield

Density (Kgm−3)	Elastic modulus (GPa)	Tangent modulus (MPa)	Poisson's ratio	Yield strength (MPa)	Ultimate strength (MPa)	Failure strain
1186	3.2	230	0.4	68	78	0.067

Bird Material Model

The actual bird is combination of flesh, blood, and bones, and it is difficult to implement the actual bird constitutive model in numerical program. Numerous researchers used different approaches to closely approximate the material response of bird. Some authors modeled the bird with elastoplastic material law along with certain failure criteria [12, 18–20], while others used equation of state approach for constitutive modeling with pressure volume behavior of water [21–24]. Bird is mostly composed of water. It behaves like a soft body and acts as a fluid on the structure during the impact. Water like hydrodynamic response by using Mie-Gruneisen equation of state (EOS) with negligible strength effects was implemented to model bird in this study. Mie-Gruneisen EOS correlates the material volumetric strength and pressure to density ratio, and it is easy to establish Gruneisen equation based on the shock Hugoniot [25]

$$P_H = \frac{\rho_o C^2 \cdot \mu (\mu + 1)}{[1 - (s - 1) \mu]^2},$$

(4)

Where

$$\mu = \frac{\rho}{\rho_o} - 1,$$

(5)

where ρ and ρ_o are the initial and instantaneous densities of material, C is the intercept, and s is linear Hugoniot slop of shock velocity (V_s) and particle velocity (V_p) relationship. Gruneisen equation describes a linear relation between shock and particle velocity, where

$$V_s = s \cdot V_p + C.$$

(6)

Gruneisen form of equation of state based on shock Hugoniot is

$$P = P_H + \Gamma \rho (E - E_H).$$

(7)

It is assumed that $\rho = {}_o\rho_o = $ constant and

$$E_H = \frac{1}{2} \frac{P_H}{\rho_o} \left(\frac{\mu}{\mu + 1} \right).$$

(8)

For $s > 1$, this formulation gives a limiting value of compression because the pressure approaches to infinity as the term $[1-(s-1)\mu = 0]$ in (4) becomes zero and the pressure therefore becomes infinite and gives maximum density of $\rho = \rho_o \, s(s-1)$. Also long before this regime is approached, the assumption of $\rho = {}_o\rho_o = $ constant is not valid. Moreover, the assumption of a linear relationship between the shock velocity Vs and the particle velocity V_p does not hold for too large compression. And at high shock strengths, some nonlinearity in this relationship is apparent. To incorporate the nonlinearity in numerical model, two linear fits to the shock velocity and particle velocity relationship were established: one holding at low shock compressions defined by $V > V_b$ and other at high shock compressions where $V < V_e$. The region between V_b and V_e is covered by a smooth interpolation between the two linear relationships [25]. The equation of state has been further improved to include a quadratic shock and particle velocity relationship

$$v_s = s_1 v_p + s_2 v_p^2 + C,$$

(9)

where s_1 and s_2 are coefficients of slop, is Gruneisen parameter, and E is internal energy. In the present numerical scheme, the water Gruneisen EOS parameters $= 0.28$, $C= 1483$, and $s= 1.75$ were used to model bird as soft projectile and other parameterswere set to zero [25]. The density of the bird was taken as 900 kg/m³ as it has been taken by most of the researchers before.

Validation of Numerical Model

Model for Windshield

In order to validate the numerical model for windshield material, FE simulations of simple uniaxial tensile test were performed. A standard dumbbell-shaped tensile test specimen with dimensions of 200 × 20 × 2.67 mm and gauge length of 50.8 mm was built. The specimen was meshed with Lagrangian solid elements and the material properties were taken as described in Section 2.1. One end of the specimen was fixed while constant velocity load was applied at the other end. The rate of loading was adjusted according to rate of change of strain. The results from FE simulations were compared with available results [17] in terms of stress strain curves at different strain rates as shown in Figure 1. The results show that the model predicts the stress-strain behavior of PMMA at different strain rates with fair accuracy.

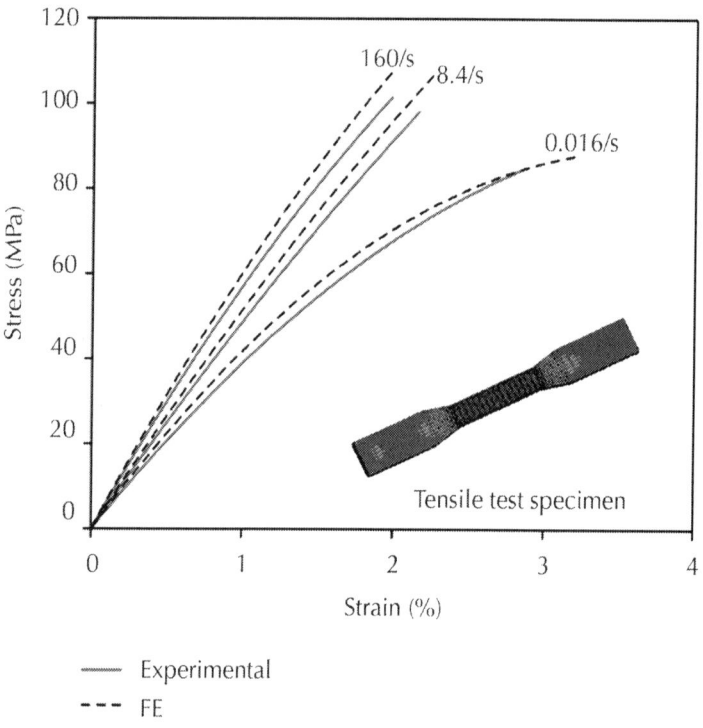

Figure 1: FE validation of windshield material model.

EOS for Bird

To validate the Gruneisen EOS implemented in FE solver, the Hugoniot and stagnation pressures were compared with theoretically obtained pressure values and Wilbeck [26] experimental results. Wilbeck experimental results carry significant importance in bird strike analysis, and many researchers used his experimental data for comparison of numerical results. For theoretical calculations, one-dimensional Hugoniot analysis was carried out in which bird impact is characterized in two stages. The first stage is initial shock called Hugoniot pressure which gives the maximum possible value of pressure during impact, and the other stage is steady state flow called stagnation pressure in which pressure stabilizes with time. The maximum pressure, that is, Hugoniot pressure, can be estimated by implying hydrodynamic impact theory and equations of conservation mass and momentum

$$\rho v_s = \rho_0 \left(v_s - v_p \right),$$

$$P_1 + \rho v_s^2 = P_2 + \rho_0 \left(v_s - v_p \right)^2, \tag{10}$$

where ρ, ρ_o, P_1, and P_2 are the initial and instantaneous density and pressure values. V_s is shock propagating velocity and Vp is particle velocity behind shock. The particle velocity V_p is assumed to be equal to the initial impact velocity V_i of bird. The Hugoniot analytical pressure can be defined as

$$P_H = P_2 - P_1 = \rho v_s v_i. \tag{11}$$

The shock velocity V_s is function of initial impact velocity V_i and can be obtained by solving the nonlinear equation [24]

$$\frac{v_s}{v_s - v_i} = (1 - \alpha) \left[\frac{v_s v_i \left(4k - 1 \right)}{C^2} \right]^{-1/(4k-1)}, \tag{12}$$

where C is velocity of sound in the material, α is porosity of material (for 10% porosity α=0.1), and value of is determined experimentally. The variation of shock velocity with initial impact velocity is shown in Figure 2.

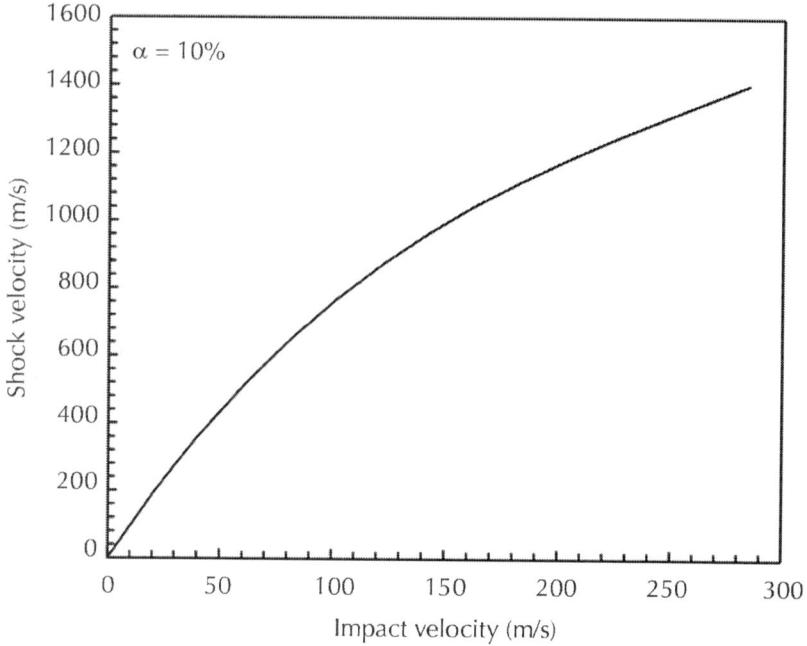

Figure 2: Shock velocity as function of impact velocity.

The steady flow pressure can be estimated by using Bernoulli relationship

$$P_s = \frac{1}{2}\rho v_i^2.$$

(13)

For FE validation, a square rigid plate of 0.5 m side was modeled in Lagrangian grid fully constrained at sides. The bird is modeled as right cylinder of 0.18 m length and 0.06 m radius with Lagrangian solid elements and material properties described in Section 2.2. In the analysis, bird with initial impact velocity of 116 m/s was impacted against rigid plate, and values for Hugoniot and stagnation pressures at central point of impact were obtained as shown in Figure 3. In order to compare the results with experimental data, the velocity of bird was taken as 116 m/s used by Wilbeck in his experiments. The results were normalized by dividing pressure with stagnation pressure and time with total impact duration and compared with experimental and analytical results.

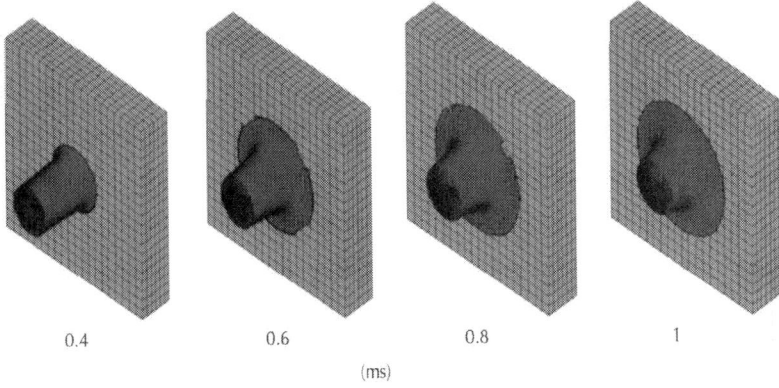

0.4   0.6   0.8   1

(ms)

Figure 3: Deformation of bird on impacting rigid plate.

From analysis, the Hugoniot pressure was predicted to have a maximum value of 102.5 MPa and the stagnation pressure is 6.5 MPa, giving normalized values of 15.7 and 1, respectively. The analytical values of Hugoniot pressure and stagnation pressure were calculated as 86.3 MPa and 6 MPa, which after normalizing gives 14.3 and 1, respectively. The comparison of FE, analytical, and experimental values of normalized pressure is shown in Figure 4. The trend of the plot is consistent with experimental results where a sudden peak pressure value was observed at initial shock and the pressure then stabilized with time. The duration of pressure decay was also in accordance with experimental results. However, the FE simulation predicted peak pressure is higher than the experimental results. The overprediction in pressure peak can be credited to bird orientation during impact, because in real bird impact experiments the initial pressure peak is highly dependent on orientation of the bird as described by Hedayati and Ziaei-Rad [24] in their work. Moreover, the Hugoniot peak pressure occurs in a very short time about 4-5 μs and the capturing of the accurate pressure peak is a challenging experimental task. The modern high frequency state of the art pressure transducers must be employed to record the accurate pressure peak during impact event, and latest experimental data in this regard is required for better comparison of FE and experimental results.

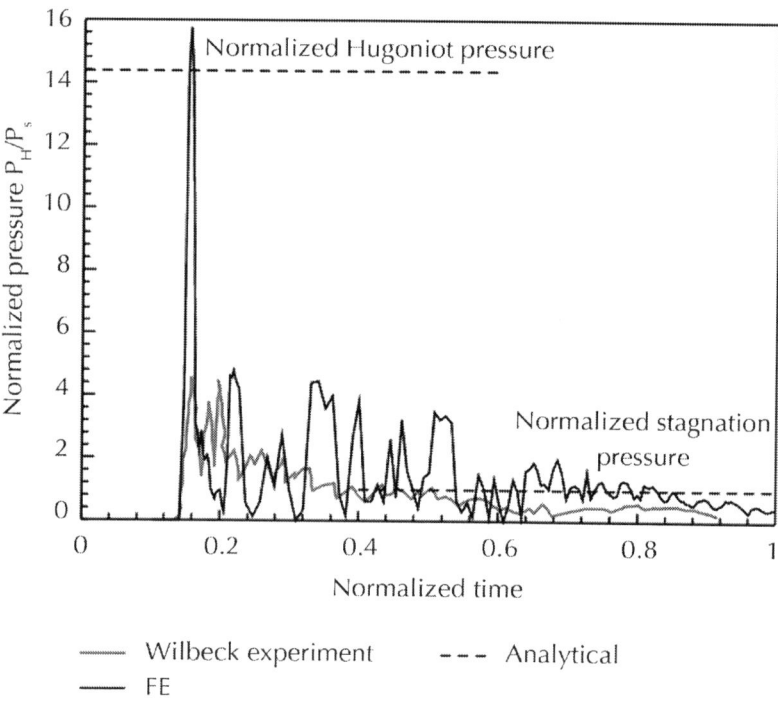

Figure 4: Comparison of results for Hugoniot pressure.

However, for the present work, the bird model was further verified by impacting the bird on a flexible metallic plate and the resultant plastic deformation after impact was determined. Welsh and Centonze [27] experimental results were considered for this purpose in which the bird with initial velocity of 146 m/s was impacted on 6.35 mm thick T6061-T6 aluminum plate and corresponding plastic deformation of the plate was measured. FE simulations were performed in accordance with experimental parameters and the residual plastic deformation δ of the plate after impact was determined as shown in Figure 5. The FE predicted deformation of 43.68 mm was in close agreement with experimentally determined value of 41.275 mm. The validated material models for bird and windshield were then used to build a full scale model for bird-windshield impact problem.

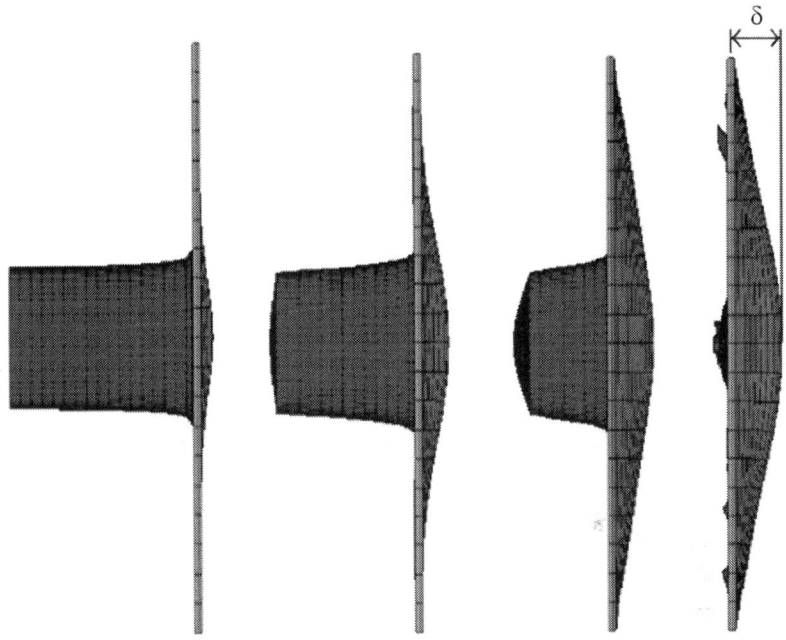

Figure 5: Bird impact on flexible aluminum plate.

3D FE Model of Impact Problem

The windshield and bird were modeled with solid elements by using Lagrangian grid as shown in Figure 6. The windshield consists of 14,400 elements with 60 elements along the curvature, 40 elements along sides, and 4 through thickness elements. More refined elements distribution is adopted around the area of impact as most of the deformation takes place at this particular impact region. The 1.8 Kg bird was modeled as right cylinder of 60 mm radius and 180 mm length. The bird was modeled as soft body with 10 elements across radius and 20 elements along length of cylinder.

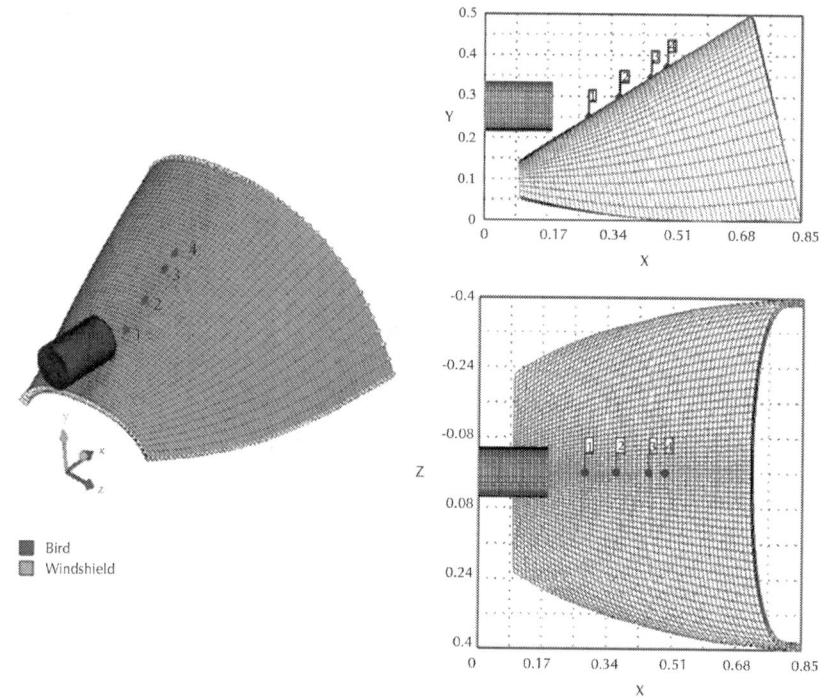

Figure 6: Finite element model of bird impact on windshield: Isometric view (Left) Side and top views (Right).

The elements in the mesh undergo severe distortion due to high rate of deformation. These distorted elements must be removed from the mesh in order to run program smoothly and to avoid reduced time step problem. The deletion of distorted elements is carried out by using geometric strain erosion model provided in AUTODYN. During the impact process, the bird highly deforms, perishes, and loses most of its mass due to elements removal process which causes inaccurate results of numerical calculations. This problem was overcome by employing the option of retaining inertia of eroded nodes in which the solver ascribes the removed cell mass to their nodal points. The contact between windshield and bird was defined by using Lagrange/Lagrange interaction option of AUTODYN. In this option, an automatic detection zone is defined around each interacting Lagrangian subgrid (independent or with itself). When a node enters into this zone, it is automatically repelled. The edges of the windshield are fully constrained to provide fixed boundary conditions. Four gage points 1, 2, 3, and 4 (Figure 6) were marked on the windshield central line to record the values of displacement, stress, and strain and to compare them with experimental results.

Experimental Confirmation

In the experimental results [11, 19], 1.8 Kg bird with velocity of 64.4 m/sec was impacted on windshield at four different locations (1, 2, 3, and 4). The displacement profiles for all the four locations were recorded during experiments. Numerical simulations were performed, and results were compared with experimental data and found in good agreement which prove the validity of numerical model as shown in Figure 7. This verified that numerical model was then further employed to investigate the effect of various parameters on windshield behavior.

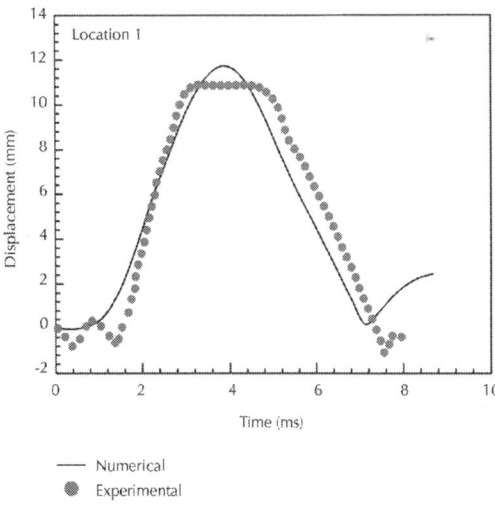

(a)

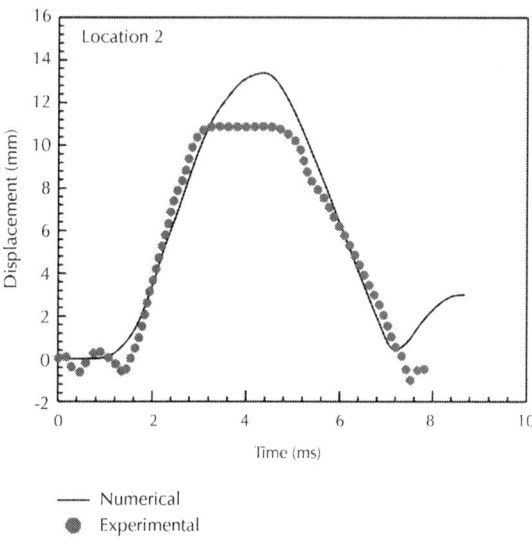

(b)

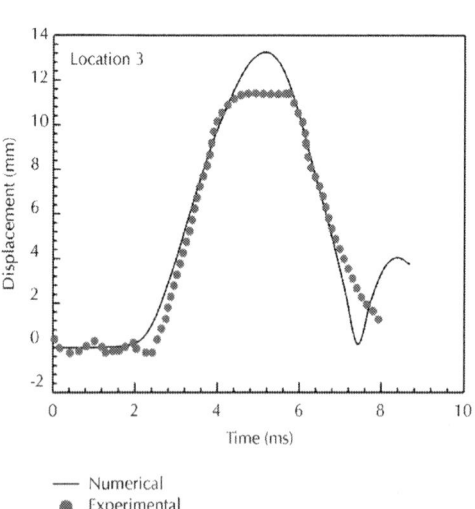

(c)

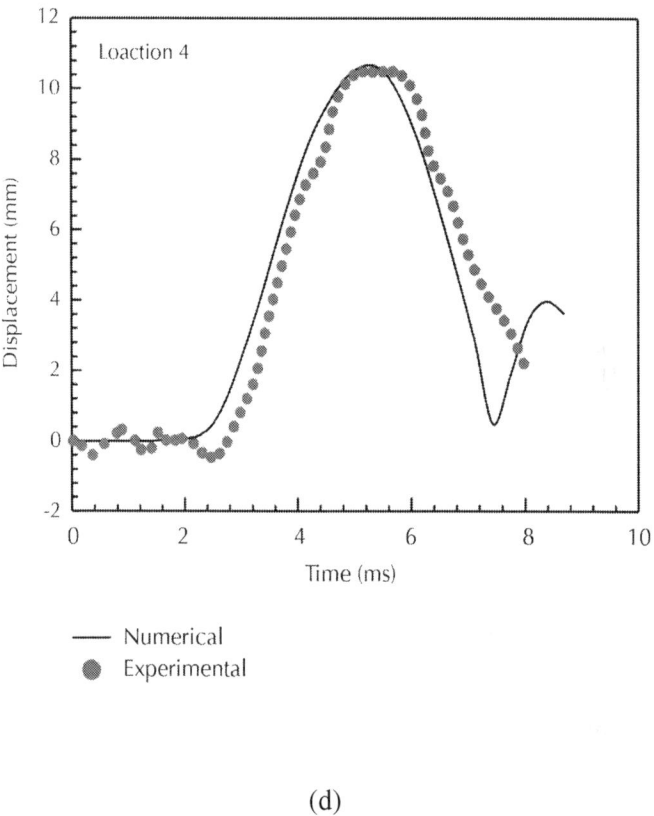

(d)

Figure 7: Displacement time plots for various locations on windshield.

SIMULATIONS

Numerical simulations were carried out to predict the dynamic response of windshield for range of impact velocities, impact angle and location of impact, and bird mass and shape. Figure 8 depicts various modes of deformation of windshield at different time intervals when impacted at location 2 with velocity of 64.4 m/sec. The windshield remains in the elastic state and keeps its original shape after impact. No sign of damage or failure was observed at this impact velocity. Also, at this velocity, the bird fails partially and slides along the surface of windshield.

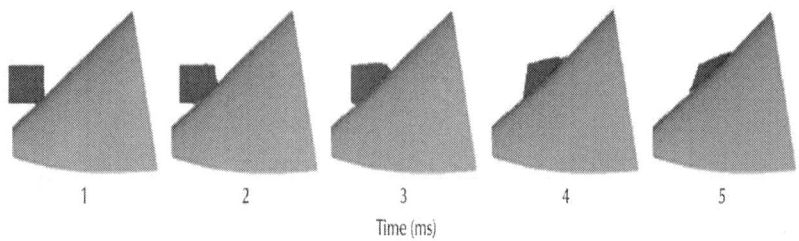

Time (ms)

Figure 8: Different stages of deformation of windshield at bird impact velocity of 64.4 m/sec.

When the velocity of bird is increased, the stresses escalate due to higher impact force and windshield tends to deform plastically. With further increase in impact velocity windshield shield reaches its elastic limit and suffers from permanent deformation leading to its complete failure. The equivalent stresses along cross-section of windshield for different velocities are shown in Figure 9. At higher velocity, the bird deforms severely and gets fragmented during sliding along the surface of windshield. Two different orientations of bird, that is, 15° and −15° from its axis, were studied to see the effect of impact angle on response of windshield and shown in Figure 10. The change of impact angle showed significant effect on normal displacement, stress, strain, and impact force. For 15° impact angle, partial bird failure occurs and whole bird body slides along the windshield surface in its way of impact.

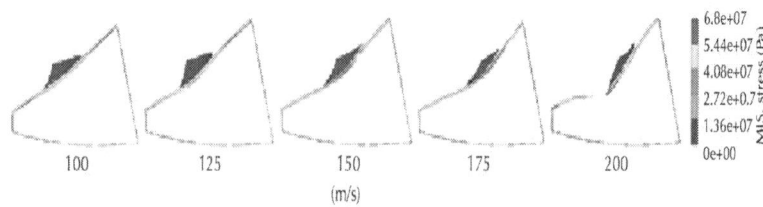

Figure 9: Impact response of windshield with increase in impact velocity.

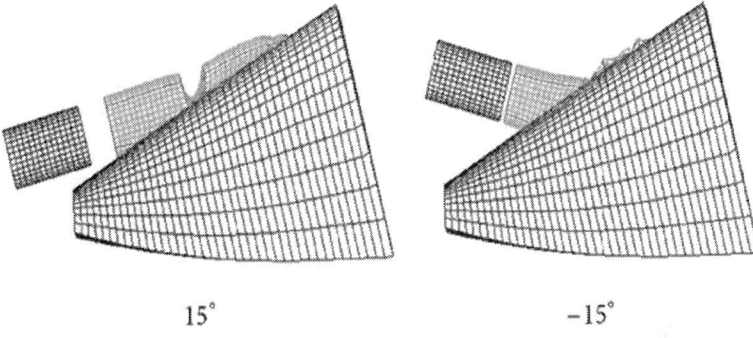

$15°$ $-15°$

Figure 10: Two different angles of impact on windshield.

At −15° angle, the bird transfers most of its energy to windshield giving higher values displacement and stress at point of impact. Most of the body of bird fails during this event.

Impact response for two additional bird geometries (hemispherical and ellipsoidal of similar length to diameter ratio) was also simulated and shown in Figure 11. For all bird shapes, the maximum length to diameter ratio and mass of the bird were taken constant. The simulations results show that impact The simulations results show that impact due to cylindrical-shaped bird produces slightly higher deformation on the windshield due to its increased contact area.

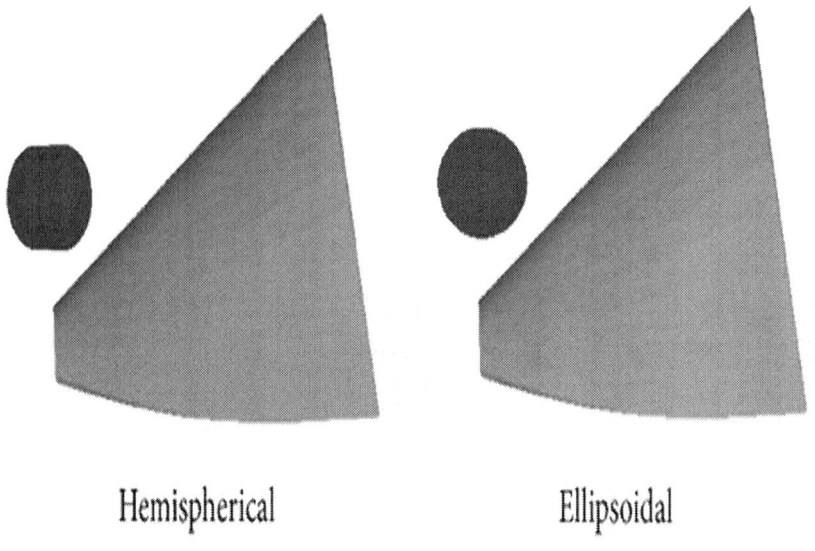

Hemispherical   Ellipsoidal

Figure 11: Response of windshield for different bird shapes.

RESULTS AND DISCUSSION

The impact response of windshield for various impact velocities was considered. It was noted that normal displacement at all gage locations increases with the increase of velocity. The time to reach maximum displacement at locations 3 and 4 increases as they lie farther from point of impact (location 2). The displacement time plots for different impact velocities at locations 2 and 4 are shown in Figures 12 and 13. Increase in velocity caused more deformation at the upper half of windshield because more of the bird mass slide and transferred more energy to the upper end. An instantaneous failure at the point of impact occurs when velocity was increased to 200 m/s.

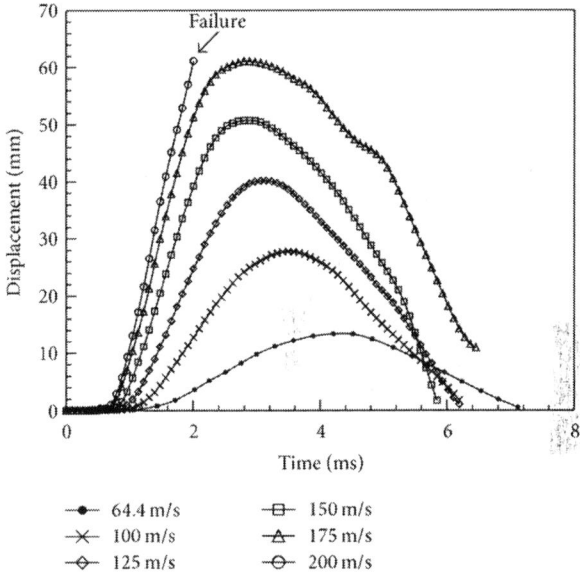

Figure 12: Displacement time plot at Location 2.

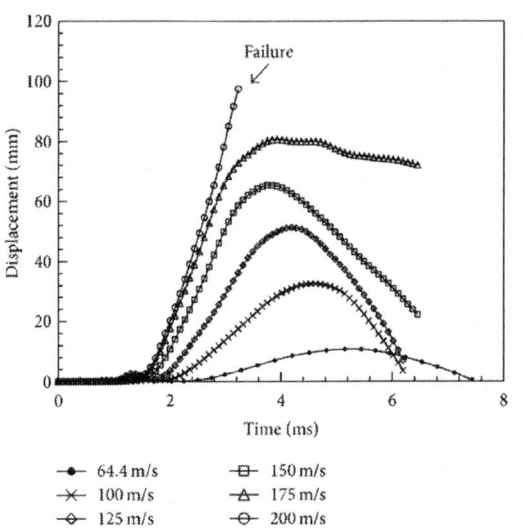

Figure 13: Displacement time plot at Location 4.

Figures 14 and 15 show the plots for equivalent stress and corresponding strain at impact point. The amplitude of stress increases with the increase of velocity; at 64.4 m/s the value of maximum stress is 30 MPa which rises up to 50 MPa at 75 m/s velocity. At 100 m/s impact velocity, the stresses in the windshield become higher than the yield stress. On further increasing the velocity to 125 m/s, the ultimate stress limit is reached and upper end of windshield fails. At 200 m/s impact velocity, the windshield failed instantly at the point of impact in 1.9 ms.

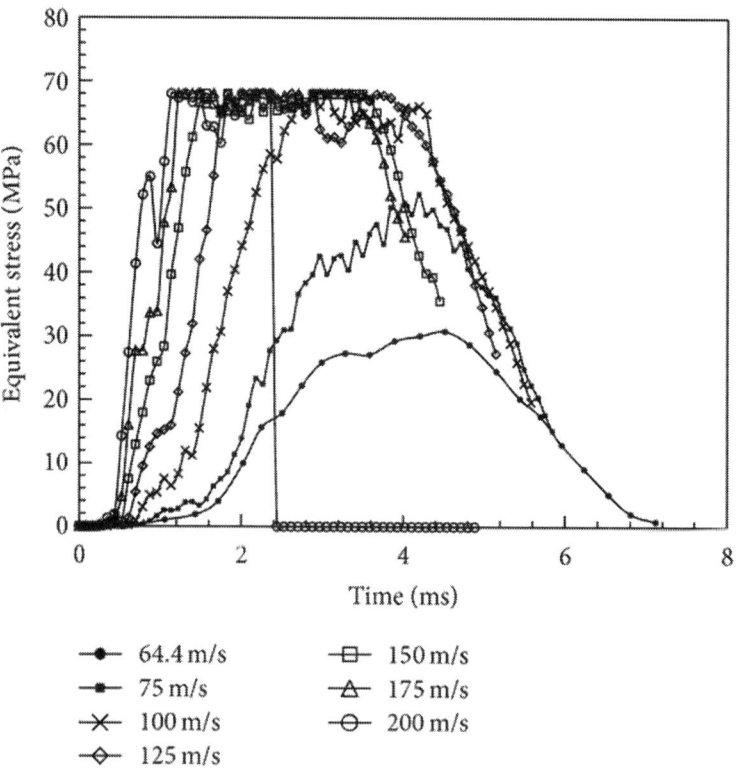

Figure 14: Equivalent stress history at Location 2.

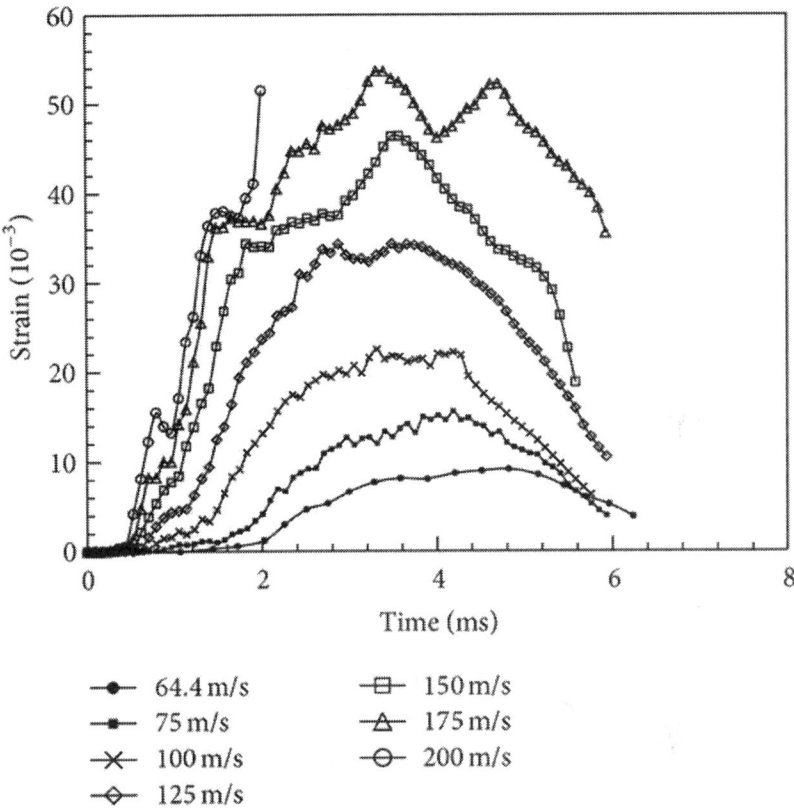

Figure 15: Effective strain history at Location 2.

Therefore, at velocity of 100 m/s and higher, the plastic deformation starts to prevail in windshield and it remains in the state of high stresses which also defines the critical impact velocity of the particular windshield. The modes of deformation and initiation of plasticity are shown in Figure 16. It can be seen that when the velocity is 100 m/s, sign of plastic deformation appears around the point of impact and upper end of windshield. The plastic deformation increases and windshield starts to fail at its upper end when the velocity is increased to 150 m/s. The further increase in velocity leads to major windshield failure at upper end and point of impact.

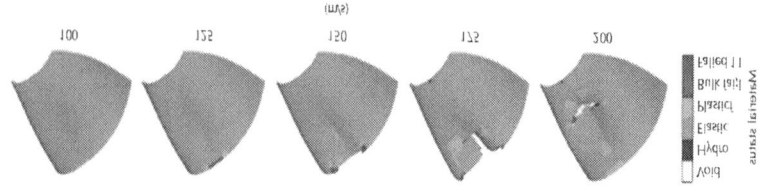

Figure 16: Deformation modes of windshield at different impact velocities.

The influence of three different bird angles at the point of impact was examined. Figure 17 shows the plots for normal displacement at 64.4 m/s impact velocity from which it can be observed that at 15° angle of impact the maximum normal displacement is 3.5 mm which increases to 13 mm for 0° and 23 mm for −15° angle. The impact force also increases sharply with the change of impact angle as shown in Figure 18. At 15° angle, the maximum impact force of 9.9 kN is recorded at 5.6 ms which rises to 48.6 kN at 2.7 ms for −15° impact angle. Almost 5 times increase in peak impact force was observed due to change in impact angle from 15° to −15°.

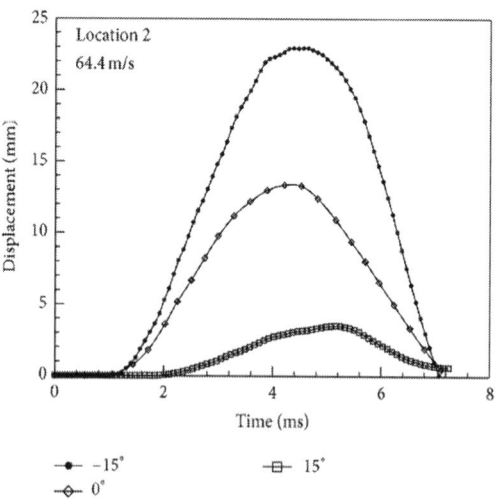

Figure 17: Displacement time plot for different impact angles.

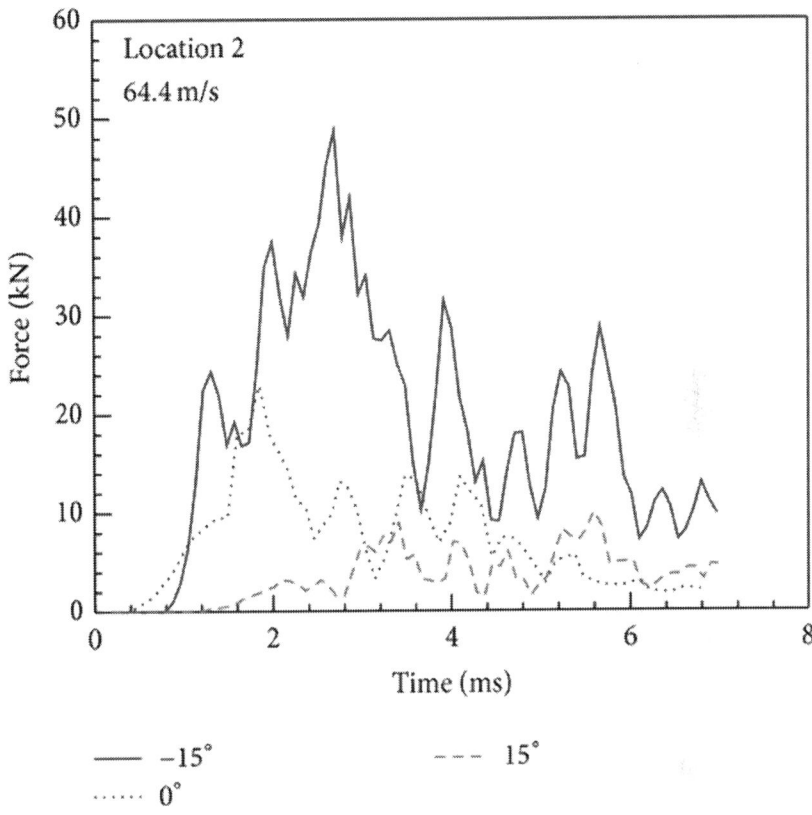

Figure 18: Impact force history for different impact angles.

The effect of impact angle on stress and strain history is shown in Figure 19. The maximum stress amplitude of 11.8 MPa and corresponding strain value of 0.00352 occur at 5.1 ms for 15° angle of impact. When the impact angle changes to −15°, the peak stress and strain value rise to 67.8 MPa and 0.0201 at 3.75 ms representing most severe impact conditions at 64.4 m/s velocity and windshield approaching its yield stress limit. Hence, impact angle is a critical parameter in determining the critical impact velocity for windshield.

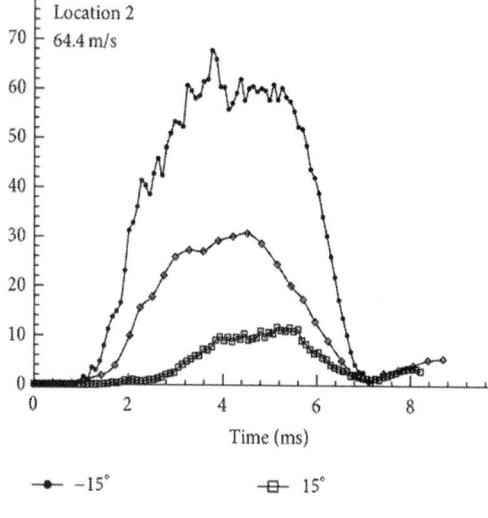

(a)

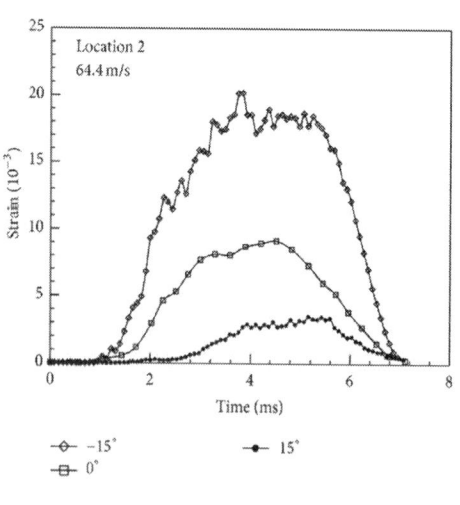

(b)

Figure 19: (a) Equivalent stress and (b) strain history during different angles of impact.

The effect of bird mass on impact response was very obvious and shown in Figure 20. With the increase in bird mass, the corresponding value of displacement and stress increases because more kinetic energy of bird is transferred to windshield. This also shows that the critical impact velocity will be different for different bird masses to produce same deformation in the windshield.

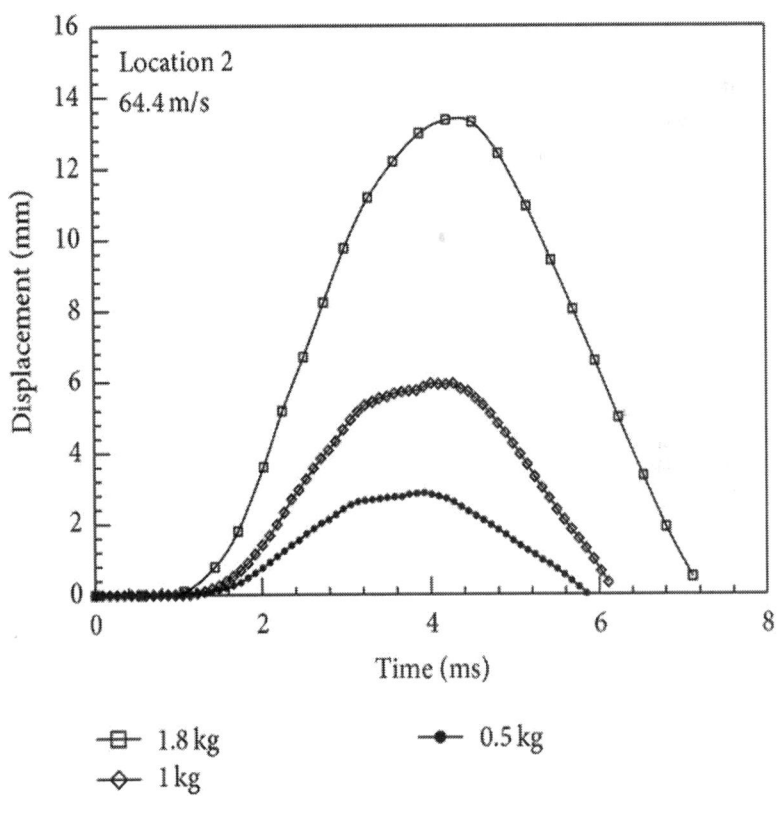

(a)

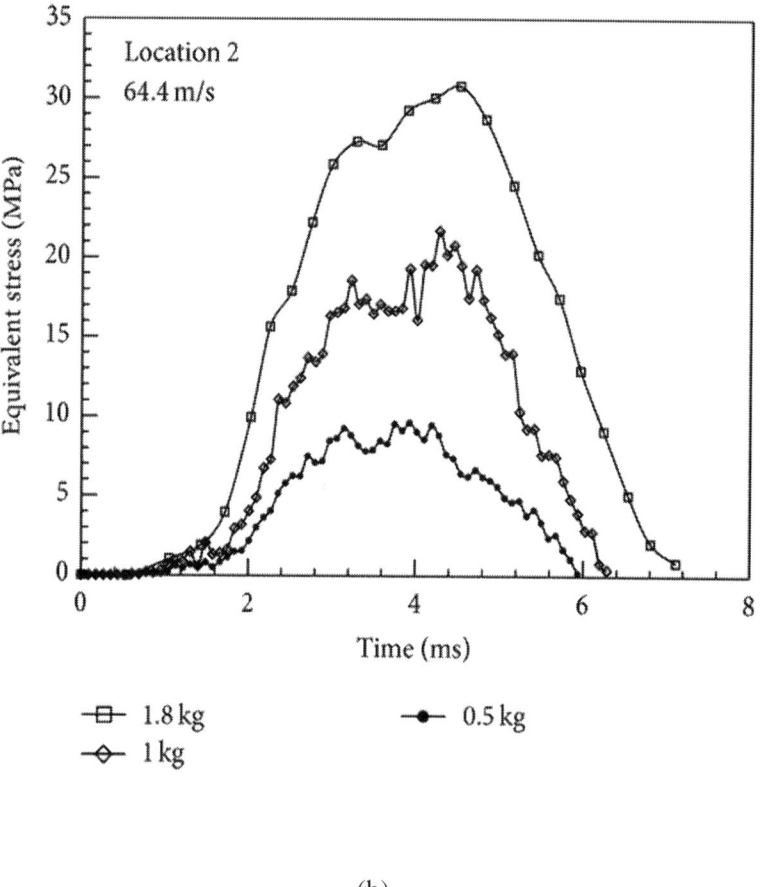

(b)

Figure 20: Effect of bird mass on (a) maximum normal displacement and (b) equivalent stress.

The response of windshield impacted by three different shape birds was studied in this work. Similar displacement trends were observed for hemispherical- and ellipsoidal-shaped birds while cylindrical-shaped bird produces higher displacement as shown in Figure 21(a). The peak values of equivalent stress for hemispherical and ellipsoidal shapes are same and higher than cylindrical shape (Figure 21(b)).

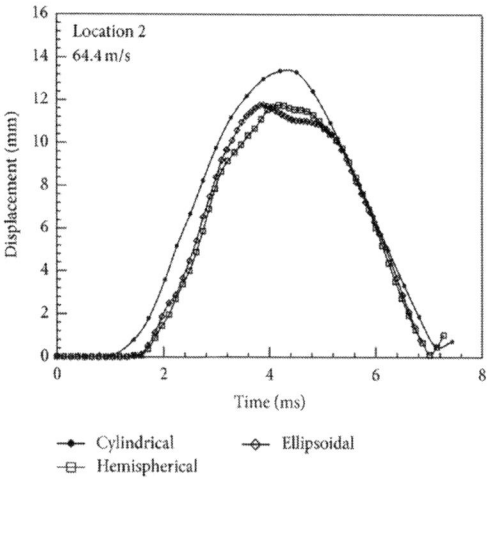

(a)

(b)

Figure 21: Effect of bird shape on (a) maximum normal displacement and (b) equivalent stress.

CONCLUSIONS

The behavior of windshield against high speed bird impact was successfully simulated and the effect of various parameters on its dynamic response was studied. Bird impact velocity was among the most critical parameter having a strong influence on dynamic response. With the increase in bird velocity, windshield tends to deform plastically beyond its yield strength which finally leads to its major failure. For a range of velocities simulated in this study, there was a limiting impact velocity at which windshield suffers permanent plastic deformation and vulnerable to fail at certain crucial locations. The rearward fixed part of windshield was considered weakest at the critical velocity. For different impact angles, the response of windshield differs greatly. With less steep angle, most of the bird slides along the surface of windshield causing less damage. Steeper angle on the other hand produces high deformation and a plastic dimple is observed at the point of impact. The bird with higher mass proved more fatal to the windshield as they impact more kinetic energy to the structure. Although the shape of the bird did not show significant effect in this study, however, the bird with smaller length to diameter ratio and higher instantaneous contact area can affect the shock pressure and peak stress level in the structure. These critical factors can be parameterized together to predict the combined effect on impact response of windshield and can provide certain guiding principles for windshield design and optimization.

ACKNOWLEDGMENTS

This work is supported by 973 Program (2012CB025904), NPU Foundation for Fundamental Research (NPU-FFR-JC201236), and Shaanxi Provincial Natural Science Foundation (2012JQ1003).

REFERENCES

1.   J. Thorpe, "Fatalities and destroyed civil aircraft due to bird strikes, 1912–2002," in Proceedings of the 26th Meeting of the International Bird Strike Committee, Warsaw, Poland, 2003.

2. S. G. Zang, C. H. Wu, R. Y. Wang, and J. R. Ma, "Bird impact dynamic response analysis for windshield," Journal of Aeronautical Materials, vol. 20, no. 4, pp. 41–45, 2000.

3. A. Samuelson and L. Sornas, "Failure analysis of aircraft windshields subjected to bird impact," inProceedings of the 15th ICAS Congress, London, UK, 1986.

4. R. R. Boroughs, "High speed bird impact analysis of the Learjet 45 windshield using DYNA3D," inProceedings of the 39th AIAA/ ASME/ASCE/AHS/ASC Structures, Structural Dynamics, and Materials Conference and Exhibit and AIAA/ASME/AHS Adaptive Structures Forum, pp. 49–59, Long Beach, Calif, USA, April 1998.

5. R. E. McCarty, "Finite element analysis of a bird-resistant monolithic stretched acrylic canopy design for the F-16A aircraft," in Proceedings of the American Institute of Aeronautics and Astronautics, Aircraft Systems and Technology Conference, Dayton, Ohio, USA, 1981.

6. R. E. McCarty, M. G. Gran, and M. J. Baruch, "MAGNA nonlinear finite element analysis of T-46 aircraft windshield bird impact," in Proceedings of the AIAA/AHS/ASEE Aircraft System Design and Technology Meeting, AIAA Paper 86-2732, Dayton, Ohio, USA, 1986.

7. F. S. Wang and Z. F. Yue, "Numerical simulation of damage and failure in aircraft windshield structure against bird strike," Materials and Design, vol. 31, no. 2, pp. 687–695, 2010.

8. F. S. Wang, Z. F. Yue, and W. Z. Yan, "Factors study influencing on numerical simulation of aircraft windshield against bird strike," Shock and Vibration, vol. 18, no. 3, pp. 407–424, 2011.

9. X. Wang, Z. Feng, F. Wang, and Z. Yue, "Dynamic response analysis of bird strike on aircraft windshield based on damage-modified nonlinear viscoelastic constitutive relation," Chinese Journal of Aeronautics, vol. 20, no. 6, pp. 511–517, 2007.

10. M. Guida, A. Grimaldi, F. Marulo, and A. Sollo, "FE study of windshield subjected to high speed bird impact," in Proceedings of the 26th International Congress of the Aeronautical Sciences (ICAS '08), 2008.

11. J. Liu, Y. L. Li, and F. Xu, "The numerical simulation of a bird-impact on an aircraft windshield by using the SPH method," Advanced Materials Research, vol. 33–37, pp. 851–856, 2008.

12. S. Zhu, M. Tong, and Y. Wang, "Experiment and numerical simulation of a full-scale aircraft windshield subjected to bird impact," in Proceedings of the 50th AIAA/ASME/ASCE/AHS/ ASC Structures, Structural Dynamics, and Materials Conference, Palm Springs, Calif, USA, 2009.

13. J. Yang, X. Cai, and C. Wu, "Experimental and FEM study of windshield subjected to high speed bird impact," Acta Mechanica Sinica, vol. 19, no. 6, pp. 543–550, 2003.

14. W. Lili, Z. Xixiong, S. Shaoqiu, G. Su, and B. Hesheng, "Impact dynamics investigation on some problems in bird strike on windshields of high speed aircrafts," Acta Aeronautica et Astronautica Sinica, vol. 12, no. 2, pp. B27–B33, 1991.

15. F. Zhou, L. Wang, and S. Hu, "A damage-modified nonlinear visco-elastic constitutive relation and failure criterion of PMMA at high strain-rates," Explosion and Shock Waves, vol. 12, no. 4, pp. 333–342, 1992.

16. A. Wang, X. Qiao, and L. Li, "Finite element method numerical simulation of bird striking multilayer windshield," Acta Aeronautica et Astronautica Sinica, Series A and B, vol. 19, pp. 446–450, 1998.

17. Z. Zhi-lin, Z. Qi-qiao, and L. Ming-xing, "Bird impact dynamic response analysis for aircraft arc windshield," Acta Aeronautica et Astronautica Sinica, vol. 9, article 018, 1992.

18. R. Doubrava and V. Strnad, "Bird strike analyses on the parts of aircraft structure," in Proceedings of the 27th Congress of the International Council of the Aeronautical Sciences, France, 2010.

19. J. Bai and Q. Sun, "On the integrated design technique of windshield against bird strike," Mechanics and Engineering, vol. 27, no. 1, pp. 14–18, 2005.

20. Y. Zhang and Y. Li, "Analysis of the anti-bird impact performance of typical beam-edge structure based on ANSYS/LS-DYNA," Advanced Materials Research, vol. 33–37, pp. 395–400, 2008

21. A. F. Johnson and M. Holzapfel, "Modelling soft body impact on composite structures," Composite Structures, vol. 61, no. 1-2, pp. 103–113, 2003.

22. J. Cheng and W. K. Binienda, "Simulation of soft projectiles impacting composite targets using an arbitrary Lagrangian-

Eulerian formulation," Journal of Aircraft, vol. 43, no. 6, pp. 1726–1731, 2006.

23. F. Stoll and R. A. Brockman, "Finite element simulation of high-speed soft-body impacts," inProceedings of the 38th AIAA/ASME/ASCE/AHS/ASC Structures, Structural Dynamics, and Materials Conference, pp. 334–344, April 1997.

24. R. Hedayati and S. Ziaei-Rad, "A new bird model and the effect of bird geometry in impacts from various orientations," Aerospace Science and Technology, 2012.

25. AUTODYN Theory manual Rev. 4.3. Century Dynamics, a subsidiary of ANSYS Inc, 2005.

26. J. Wilbeck, "Impact behavior of low strength projectiles," Report AFML-TR- 77-134, Air Force Materials Laboratory, 1977.

27. C. J. Welsh and V. Centonze, "Aircraft transparency testing artificial birds," Report AEDC-TR-86-2, US Air Force, 1986.

Chapter 3

Microtomographic Analysis of Impact Damage in FRP Composite Laminates: A Comparative Study

M. Alemi-Ardakani[1], A. S. Milani[1], S. Yannacopoulos[1],
L. Bichler[1], D. Trudel-Boucher[2], G. Shokouhi[3], and H.
Borazghi[3]

[1]School of Engineering, University of British Columbia, Kelowna,
Canada V1V 1V7

[2]Industrial Materials Institute, National Research Council, Boucherville,
Canada J4B 6Y4

[3]AS Composite Inc., Pointe-Claire, Canada H9R 4L6

ABSTRACT

With the advancement of testing tools, the ability to characterize
mechanical properties of fiber reinforced polymer (FRP) composites
under extreme loading scenarios has allowed designers to use these

materials in high-level applications more confidently. Conventionally, impact characterization of composite materials is studied via nondestructive techniques such as ultrasonic C-scanning, infrared thermography, X-ray, and acoustography. None of these techniques, however, enable 3D microscale visualization of the damage at different layers of composite laminates. In this paper, a 3D microtomographic technique has been employed to visualize and compare impact damage modes in a set of thermoplastic laminates. The test samples were made of commingled polypropylene (PP) and glass fibers with two different architectures, including the plain woven and unidirectional. Impact testing using a drop-weight tower, followed by postimpact four-point flexural testing and nondestructive tomographic analysis demonstrated a close relationship between the type of fibre architecture and the induced impact damage mechanisms and their extensions.

INTRODUCTION

During experimental analysis of impact behaviour of FRP composites, it is common to use nondestructive/destructive detection methods to investigate the induced damage modes and their extension in test samples. Different nondestructive methods have been used in the literature, from simple visual methods [1–5] to more complex thermal- or electrical-based [6–9] methods, ultrasonic C-scanning [10–13], and X-ray imaging [14–16]. Each method has its own advantages and disadvantages and may be suitable for a particular application/material type. Nevertheless, a common limitation of these methods is that they are generally unable to give a full 3D image of the interior part of the material, hence making it difficult to provide complete information regarding the location and extent of different damage modes such as matrix cracking, fiber breakage, fiber pull-out, fiber-matrix debonding, and delamination. On the other extreme, the destructive methods have been of less desire for sensitive applications as they can be the source of additional damage in the impacted zone of structures such as fiber breakage, fiber pull-out, or delamination growth. The previous shortcomings can be well addressed by using today's advanced X-ray microtomography techniques (XMTs), which is the main focus of this paper. Namely, the present work aims at a detailed comparison of damage state in impacted woven fabric and unidirectional thermoplastic laminates via XMT, thereby arriving at a correlation between the observed damage distributions and the underlying reinforcement type.

Historical Background

X-ray microtomography technique (XMT) is known as a nondestructive technique for 3D microstructure reconstruction and visualization of the interior parts of objects with a resolution in the order of micrometers. Johann Radon, a Czech mathematician, was the first scientist who conceived a mathematical solution for the reconstruction of X-ray images in 1917 [17]. Allan Cormack, a South African physicist, continued the previous work and developed an algorithm for the geometrical reconstruction problem at Tufts University in 1964. Following this, Godfrey Hounsfield built the first CT (computed tomography) scanner at EMI Research Labs in the UK in 1972. It is worth mentioning that Cormack and Hounsfield received Nobel Prize in 1979 because of their contributions in building the first CT scanner and its effect on medical imaging applications [17]. Figure 1 shows a schematic of the macrotomography technique that today is used in medical examinations. Medical CT scanners use a point source X-ray and an array of detectors. The patient body is inserted into the machine chamber. At the same time X-ray source and detectors rotate around the body and collect the X-ray images, that is, helical body scanning.

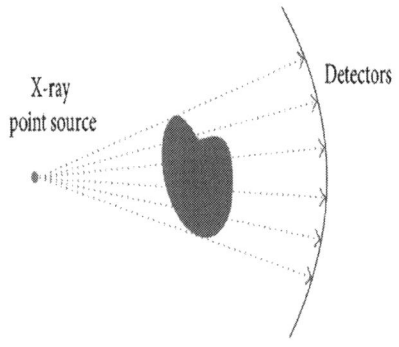

(a)

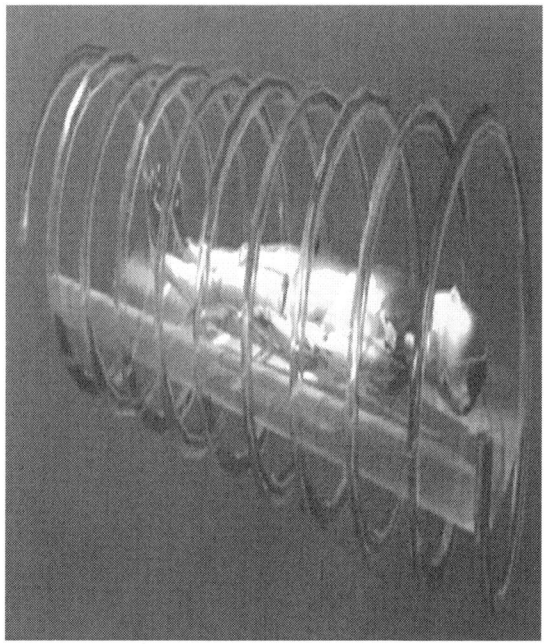

(b)

Figure 1: (a) Schematic of macrotomography used in medical examinations; (b) helical body scanning [17].

Because of the sensitivity of human body to high radiation exposure, the energy and dosage of X-rays in these machines are set to be low and as a result the ensuing image resolutions are often low [18]. This limitation led Elliot and Dover in 1982 to build a more precise machine with higher exposure capability and image resolution (12 μm) for industrial applications and microanalyses [19]. Another difference between the industrial microtomography machines (XMT) and the medical CT scanners is that the X-ray source and detectors in XMT machines are stationary and the sample rotates. Depending on the need, one can set the machine to take several thousands of scans in a complete rotation of the sample between 0° and 360°. Subsequently, postprocessing software is used to reconstruct the 3D image of the sample which contains all geometrical information of the interior microstructure.

Example of XMT for Composites

Before presenting the conducted case study, let us illustrate a general example of an XMT image (obtained by Xradia microXCT-400 machine) as compared to an image obtained from the same sample through a destructive method. Namely, an impacted composite sample was cut with a slow speed diamond saw (Figure 2) and the cross-section of the impacted zone was examined by an optical microscope (Figure 3). In the nondestructive counterpart of this analysis, the specimen needed no physical cutting and the XMT image (Figure 3) shows a slice (virtual cut) of the 3D image of the material microstructure in the midplane. The comparison of the two images shows that microtomography has captured the interior damages reasonably well. Slight differences between these images can be due to the damage induced during the cutting process in the destructive method (microscopy) such as fiber breakage, fiber pull-out, cracking, compressing or opening the delaminated layers, which in turn implies an advantage of using tomography as a nondestructive method. Additionally, in the destructive method the cut sample may not be used for further investigations at different planes, whereas in the XMT the virtual cutting plane can be moved over the sample to scrutinize the microstructure in arbitrary sections. Figures 4(a) to 4(d) show the trend (histogram) of such interactive analysis for four different cutting planes (namely, at different distances from the specimen center). Each XMT slice has four subimages (top, front, and left views). The cutting planes are shown by red, blue, and green lines. Figure 4(a) reveals the damaged cross-sections when the top and left cutting planes (blue and red lines) are far from the impact center as noted in the front view. Figure 4(b) shows the tomography slice when the top cutting plane (blue line) was placed near the impact center. In Figure 4(c), the left cutting plane (red line) was moved towards the impact center and in Figure 4(d) the left cutting plane was almost at the center where most of the damage is noticed from the side view.

Figure 2: A high precision saw cutting the composite sample in the mid plane where the damage zone is present.

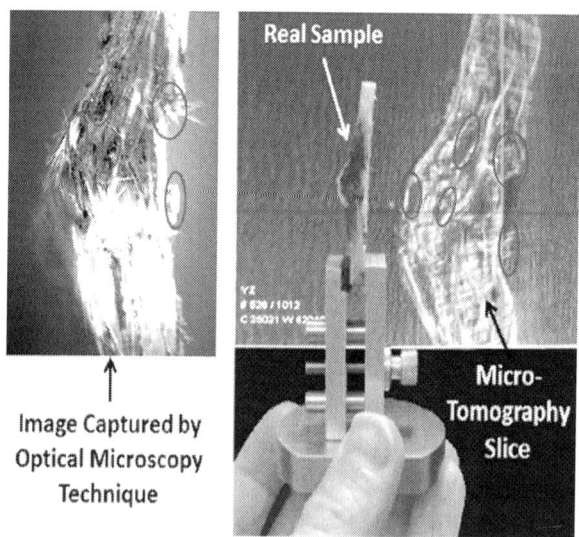

Figure 3: Comparison between the real impacted sample and the images obtained from nondestructive microtomography and destructive optical microscopy; red circles are to show comparable damage zones captured by the two methods.

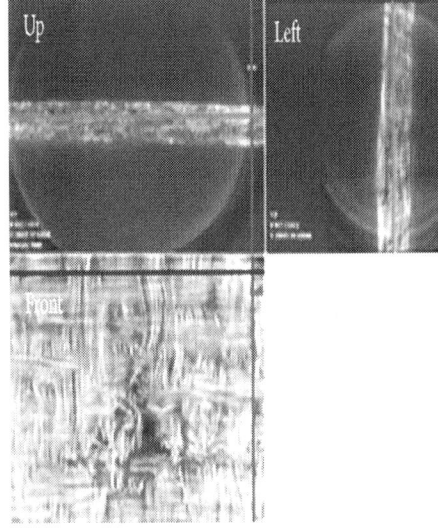

(a)

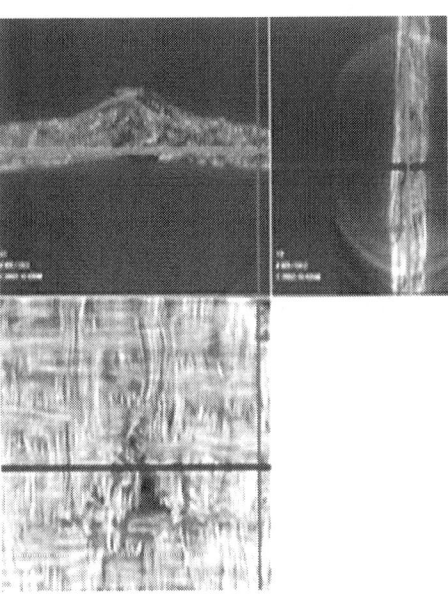

(b)

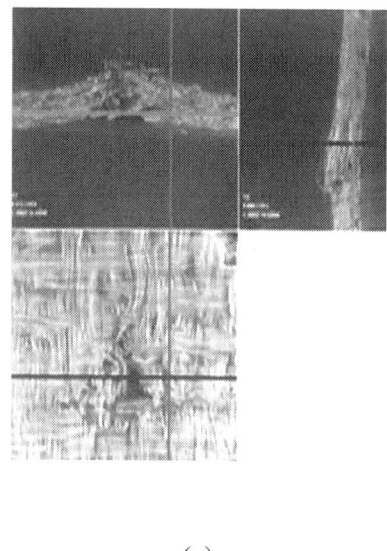

(c)

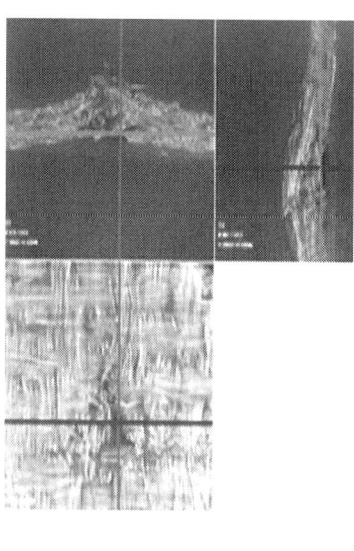

(d)

Figure 4: Example of the histogram of XMT analysis in an impacted laminate; the varying severity of fiber distortion, matrix cracking, and delamination can be noticed depending on the distance from the impact center.

CASE STUDY

Sample Preparation

Two sets of test samples were prepared using vacuum bagging to laminate 12 layers of polypropylene/E-glass preform (with a fiber volume fraction of 60%–70%) using two different reinforcement patterns including the plain woven (PW) and unidirectional (UD). The laminates' size was chosen for impact testing based on ASTM D7136 [20] with a rectangular shape (150 × 100 mm) and the total thickness of 6 mm.

Impact and Postimpact Flexural Testing

Each composite laminate type (PW or UD) was tested under a drop-weight tower (Figure 5(a)). Impact tests were conducted using Dynatup Model 8200 impact machine with a hemisphere projectile of 1 inch in diameter and 12.26 kg mass. Each test was repeated twice and all four sides of the specimens were completely clamped during the impact event. The impact energy was kept constant at 200 J. Force history during the impact event was collected by a load cell, a quartz piezoelectric force sensor, mounted on the impactor. The acceleration of impactor as a function of time, a(t) , was calculated by Newton's second law of motion (1) from the collected force history F(t) and the impactor mass :m

$$a\left(t\right) = \frac{F\left(t\right)}{m}.$$

(1)

(a)

(b)

Figure 5: Set-up used during (a) the drop-weight impact testing and (b) post-impact four-point bending.

The velocity v(t) and displacement of impactor x(t) were found by numerical integration as

$$v(t) = v_0 + gt - \int_0^t \frac{F(t)}{m} dt,$$

$$x(t) = \int_0^t v(t) dt,$$

(2)

where v_0 is the velocity of impactor at the time of hitting the sample measured by an infrared velocity detector; see [21] for more details of drop-weight test kinematics. After impact testing, a postimpact four-point flexural experiment (Figure 5(b)) was conducted on each specimen. The motivation was to study the postimpact resistance of the impacted composite laminates for their potential application, for example, as a highway guardrail between inspection/repair intervals and also to find the deterioration of their effective mechanical properties due to the impact event. All the results presented in the next sections are normalised with respect to the fiber volume fraction.

Results of the Impact and Postimpact Bending Tests

Figures 6(a) and 6(b) show the average contact force and displacement of projectile from repeats of the test. In comparison to UD samples, Figure 6(a) suggests that PW has exerted more force to the impactor. The energy has been calculated via

$$E(t) = \int_0^{x(t)} F(t) dx$$

(3)

where F is the reaction (contact) force and x is the impactor displacement. Figure 6(c) shows the average energy of impactor for the two experiments. Subtraction of the energy of impactor at the time of hitting the sample (200J) from that at the rebounce indicates the dissipated energy due to permanent damage in the material. This energy is represented by the area trapped between the penetration and rebound curves in Figure6(b) or the final flat energy level in Figure 6(c) after about 8 ms. According to these diagrams, UD laminates have absorbed more energy than PW laminates. Hence, it may be concluded that the absorbed energy has been decreased by increasing the reinforcement waviness from unidirectional to plain weave pattern, given comparable laminate thicknesses and fiber contents.

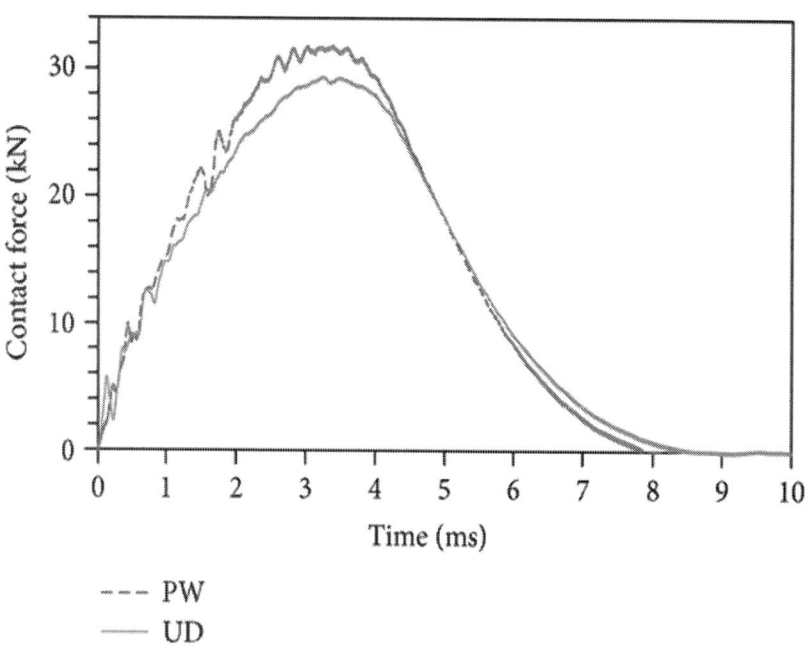

PW

UD

(a)

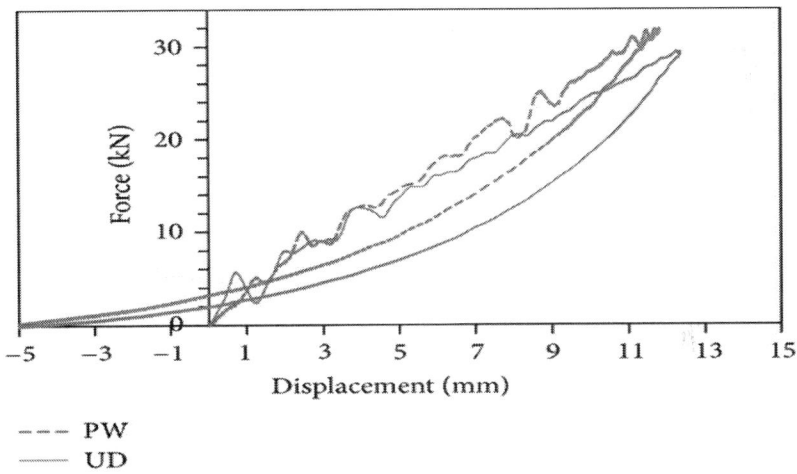

(b)

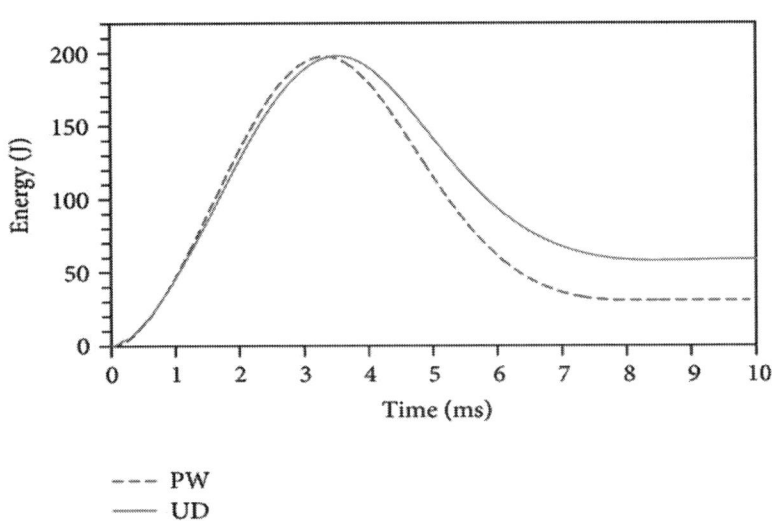

(c)

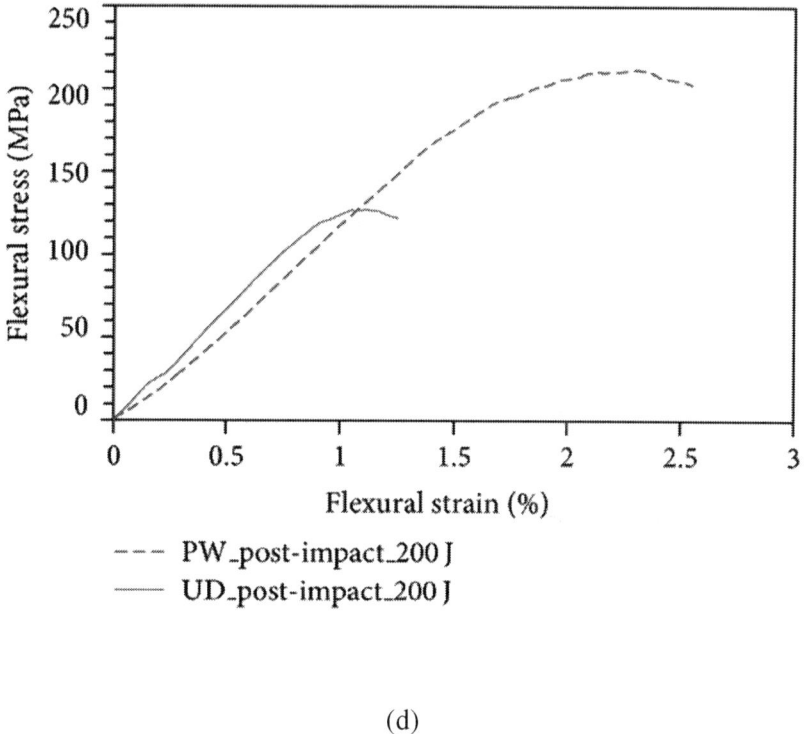

(d)

Figure 6: The average response curves of PW and UD samples subjected to 200 J impact loading; (a) impactor force versus time, (b) force versus displacement, (c) energy versus time, and (d) the postimpact flexural test results.

Figure 6(d) shows the average results of flexural testing for impacted samples. It confirms that the impacted plain woven composite has withstood postimpact bending forces much better than the impacted unidirectional composites. For comparison purposes, the four-point flexural testing was also performed on PW and UD healthy samples (i.e., before impact damage). Accordingly, Figure 8 indicated that the deterioration percent of ultimate flexural strength due to impact is 19% for the PW material and 32% for the UD material. This result is in agreement with the energy results in Figure 6(c): the more the absorbed energy by the material, the higher the deterioration of effective mechanical properties of the sample after the impact. Hence, we can conclude that UD samples have been damaged more severely than PW samples under impact. However a question would then be why is the visible (exterior) damage in PW samples much more apparent than UD samples as illustrated in Figure 7? XMT technique was employed to answer this question as it can illustrate the interior damage of the samples.

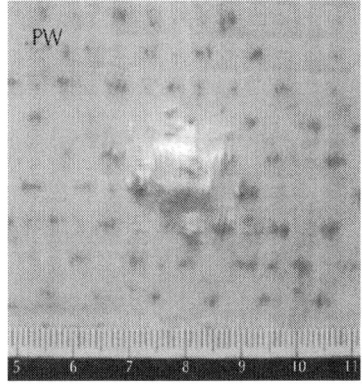

(a)

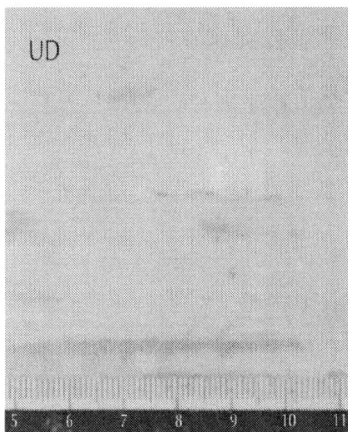

(b)

Figure 7: Rear face of the PW and UD samples subjected to 200J impact energy.

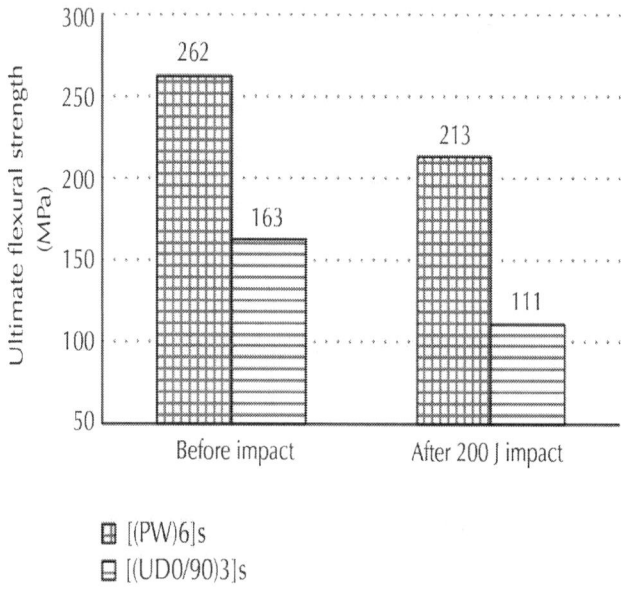

Figure 8: Ultimate flexural strength comparisons before and after impact tests on UD and PW laminates.

XMT Results

As addressed earlier, X-ray microtomography tests were conducted using Xradia microXCT-400 machine with sample dimensions of 6 × 40 × 120 mm. Table 1 shows the acquisition parameters and the test set-up used during tomography. Images obtained by this technique comprised 1024 × 1024 pixels of 33.57 μm.

Table 1: Tomography acquisition parameters used during imaging

X-ray source	Detector	Tomography setup
Power = 10 watt Voltage = 62 kV Current = 155 μA	Magnification = 0.39 X Filter: No	No. of radiographs = 630 images Angle of rotation: −110° to 110° Illumination time = 1 s per radiograph

Figures 9(a) and 10(a) show the XMT images of PW and UD laminates at selected cross-sections. Figures 9(b) and 10(b) show a 25 mm × 10 mm window cropped from the top and left cross-section views of PW and UD specimens. Fiber layers are also marked in these images. Figures 9(c) and 10(c) represent Figures 9(b) and 10(b) after image processing using Buehler Omnimet 9.5 software. The image processing enabled measuring the damage areas quantitatively. The green and dark red regions in Figures 9(c) and 10(c) indicate the healthy and damaged regions, respectively. Figure 9(c) suggests the presence of several delamination sites, matrix crushing, and separations (branching) of fiber bundles within inner layers. This view also shows fiber breakage of two layers close to the impact center as well as a large delamination between the third and fourth layers.

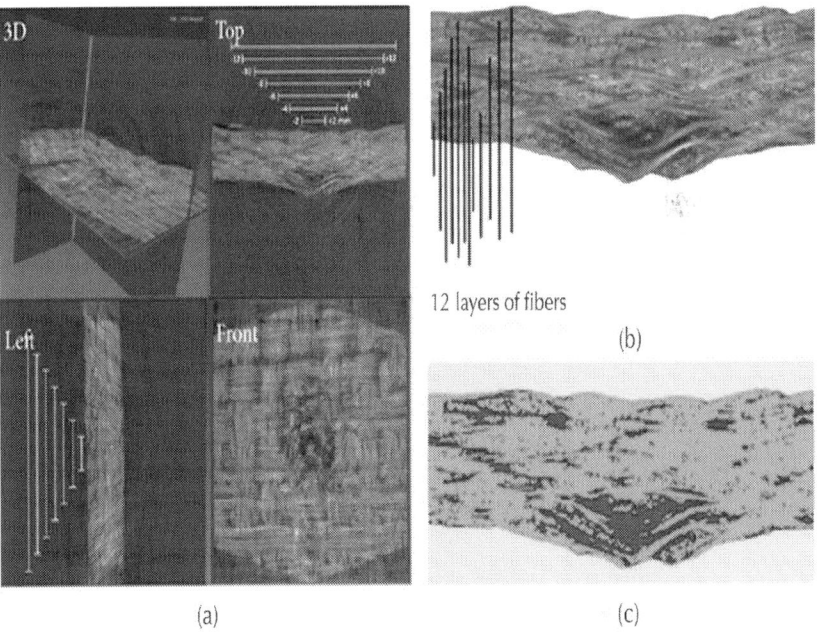

(a)   (c)

Figure 9: (a) XMT image of the impacted PW laminate (top view at the impact center, left view at 10 mm from the impact center, and front view close to the rear side of impact), (b) enlarged top view within a cropped window of 25 mm × 10 mm, and (c) the processed image of top view at the impact center (for subsequent damage quantification purposes).

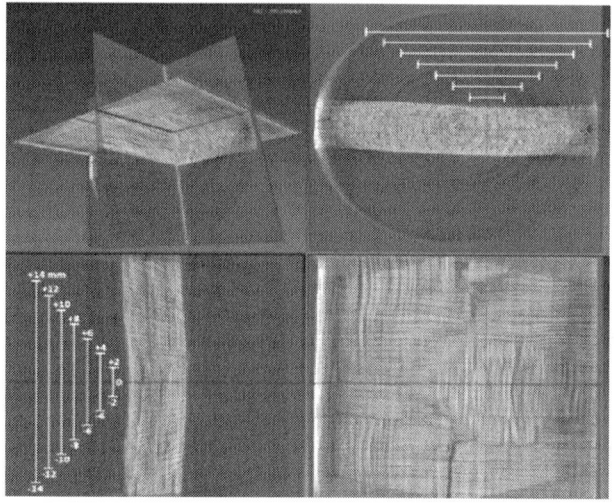

(a)

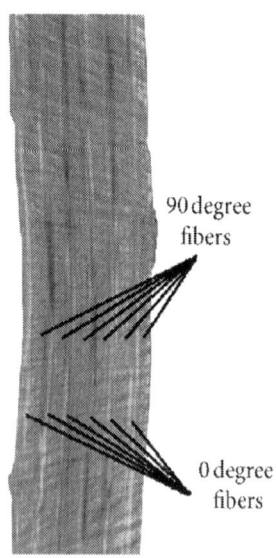

90 degree
fibers

0 degree
fibers

(b)

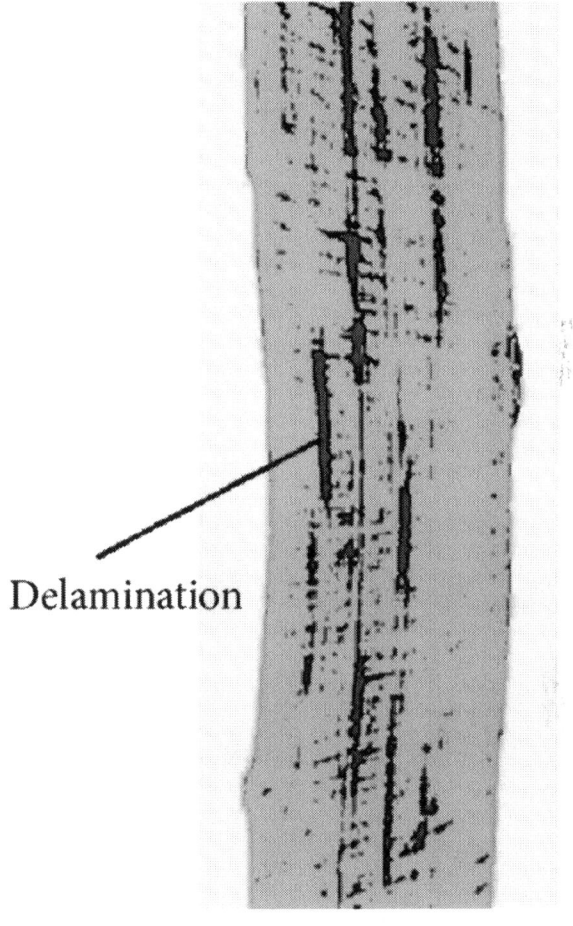

Delamination

(c)

Figure 10: (a) XMT image of impacted UD laminate (top view at the impact center, left view at 10 mm to the impact center, and front view close to the rear side of impact), (b) enlarged left view within a cropped window of 25 mm × 10 mm, and (c) the processed image of left view at the impact center.

A set of virtual rulers were placed in the top and left views of both Figures 9(a) and 10(a) with the total lengths of 28, 24,…, 4 mm. These rulers were used as indicators for subsequent image analyses to cut the 3D XMT images from −14 mm to +14 mm distance from the impact center with a spacing of 2 mm. Images obtained from these cuts on one side of the impact center are presented in histogram forms (Figures 11 and 12). The useful length of field of view in collected tomography images was considered to be 20 mm (to avoid edge effects that deteriorate the image resolution); hence the results in Figures 11 and 12 were included up to 10 mm (on one side) from the impact center.

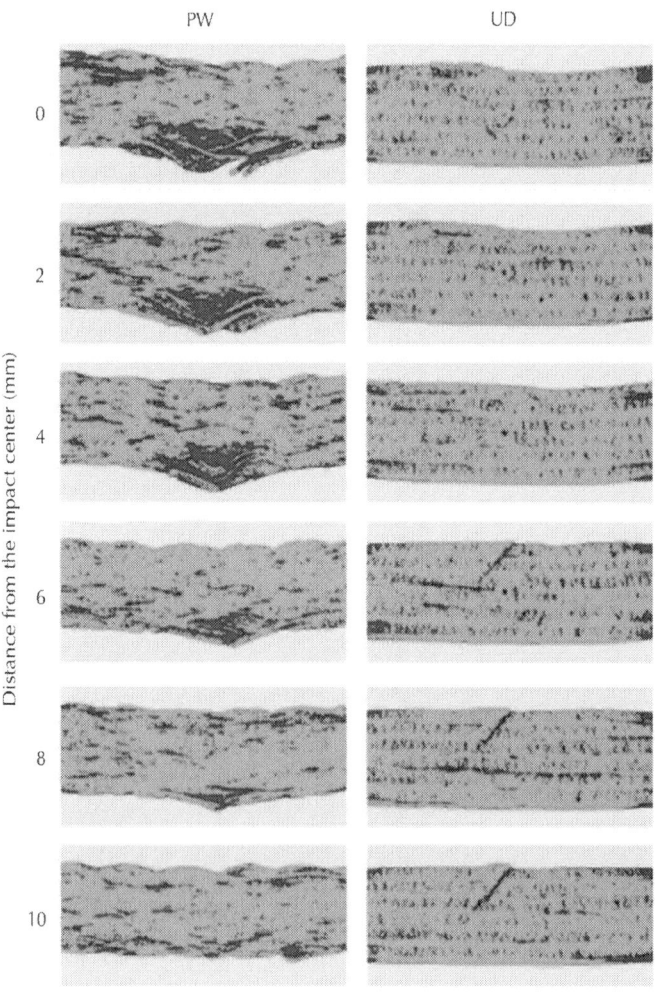

Figure 11: XMT top view histogram of the impacted PW and UD laminates.

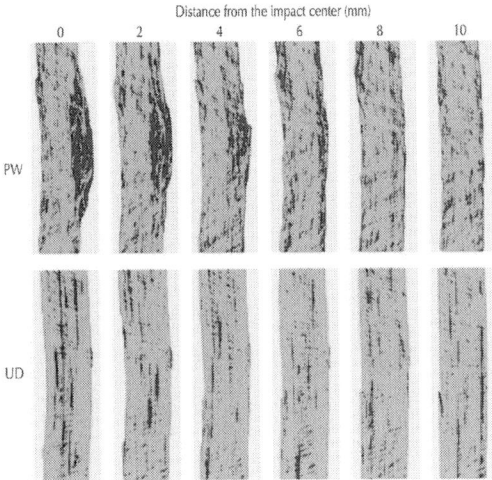

Figure 12: XMT left view histogram of the impacted PW and UD laminates.

Comparing results in Figures 11 and 12, it is first noticed that the states of damage at the top and left cross-sections are not generally identical, given the same cut distance from the impact center (specially for the UD sample). This is most likely because of the nonsymmetric impact boundary condition during the drop-weight tests due to the nonsquare shape of the fixture (130 × 80 mm). Expectedly, cracks and delaminations have been propagated longer in the direction with larger specimen dimension, which lays on the left view of tomography images. As the cutting plane goes farther from the impact center, we notice that the PW sample appears to be more and more undamaged (comparable to the healthy state).

For UD laminates, from Figure 10(a), no severe local damage is observed under the impact center. There were, however, well-distributed small dark regions (dots) on the top view (see Figure 11 for results). Each of these dark regions would correspond to a delamination which can be traced in the corresponding left view in Figure12. It was interesting that, in contrast to the PW laminate, if we go far from the impact center (up to 10 mm which was the maximum useful field of view), there still exits evidence of some locally delaminated zones in the UD laminate and their intensity does not decrease rapidly. This means that the extent of damage in the UD laminate in the form of several microdelaminations would be higher than that in the PW sample. Comparing the top view histograms of UD and PW laminates in Figure 11, another main difference between the two impacted materials is revealed: a very large through-thickness crack and fiber breakage have occurred in the UD laminate starting from the impacted face of the sample (marked with a white arrow in Figure 13). In fact, the calculated larger magnitude of absorbed energy in the UD laminate, 57.218J versus 36.2J for the PW sample, also shown in Figure 6(c), could be linked to this large through-thickness crack and fiber breakage in addition to the aforementioned distributed local delaminations across the sample. It should be added that a similar crack was visible in all test repeats of the UD material. Relating to the reinforcement architecture, the high waviness in the plain woven laminates would act as a barrier against impact pulse. On the other hand, flat UD fibers have allowed the impact wave to propagate from the center to the structure more easily without a large local damage under impactor.

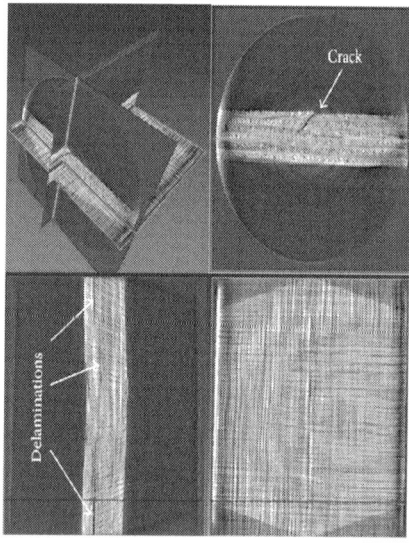

Figure 13: A cross-section of the impacted UD laminate showing a large through-thickness crack and clear delamination sites. A similar crack was observed in all test repeats for this material.

Figure 14 shows the area fraction of damaged regions (dark zones) obtained quantitatively from processed images in Figures 11 and 12. Each data point in Figure 14 has been calculated from the average response of the two cross-sections located symmetrically with respect to the impact center. According to the observed trends, the inner damaged area of PW samples decreases linearly by the distance from the impact center. Interestingly, in contrast to the PW sample, the damage faction of the UD sample has not varied notably by the distance from the impact center—it is nearly constant after 2 mm across the sample within the given field of view. This result, in turn, confirms that the damage distribution has been more uniform in the UD sample. Also Figure 14suggests that the damage fraction of UD samples has been overall lower than PW samples. On the other hand, as discussed in the previous sections, the deterioration of effective mechanical properties from the healthy to impacted samples has been more severe in the UD material (also this material has absorbed more energy as shown in Figure 6(c)). This means that for impact damage analysis and its linkage to the residual mechanical properties in the samples, next to the damaged area, one should look into other associated parameters. One of these key parameters is the corresponding damage mode to each damaged area. Although there is no evidence of severe local damage under impact centre in UD samples, the very large through-thickness brittle crack and the associated fiber breakage mode, along with the distrusted delaminations, have played a significant role in the absorption of impact energy in this material.

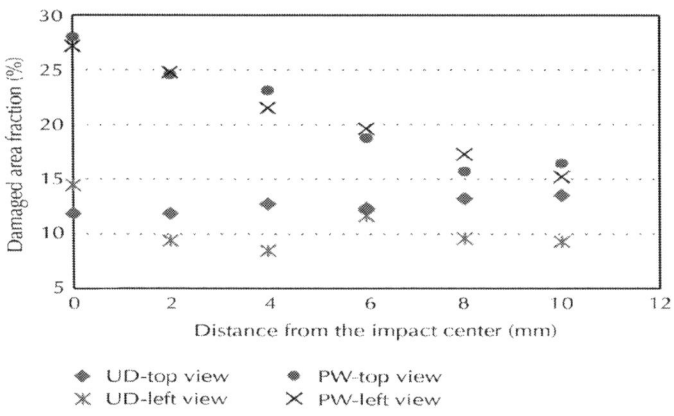

Figure 14: Area fraction of damaged zones in the impacted UD and PW samples, as function of distance from the impact center.

CONCLUSIONS

PP/glass thermoplastic laminate samples were made using (i) unidirectional fibers (UD) and (ii) plain woven (PW) fabrics and subjected to 200 J impact energy as well as postimpact four-point bending. UD specimens absorbed (dissipated) more energy than the PW laminates. This was despite the fact the UD samples showed no or very little visible damage area in the outer faces. X-ray microtomography technique (XMT) was used to investigate the damage and its distribution inside the specimens. XMT analysis showed that the impact energy has been absorbed to create a severe local damage under the impact center of PW laminates, whereas well-distributed delamination zones were found across the unidirectional laminates even far from the impact center. The reason would be that unidirectional fibers allow the impact wave to propagate more easily through the structure, whereas the waviness of woven fabrics can act as a barrier for damage propagation. A large through-thickness crack was also seen inside the UD sample, which has broken 6 out of 12 layers of the laminate. In summary, this case study suggests that the rear side visible damage in impacted FRP laminates cannot represent the entire damage extension and the associated loss of effective mechanical properties (here identified through postimpact flexural testing). Microcracks and distributed local delamination sites "inside" the samples can significantly contribute to the dissipation of impact energy. A powerful nondestructive inspection method such as XMT can be used to visualize and quantify the damage state and its extent inside the specimens in 3D. Some clear differences were seen between damage states inside the impacted UD and PW laminates and suggested that reinforcement selection should be made with ultimate care depending on the objectives of a given impact application. A worthwhile future study may be the XMT analysis of hybrid laminates with varying thickness and reinforcement architecture subject to different levels of impact energy.

ACKNOWLEDGMENTS

The authors wish to acknowledge financial support from the Natural Sciences and Engineering Research Council (NSERC) of Canada. Comments and suggestions of the anonymous reviewers are also greatly acknowledged.

REFERENCES

1.   M. A. Ardakani, A. A. Khatibi, and H. Parsaiyan, "An experimental study on the impact resistance of glass-fiber-reinforced aluminum (Glare) laminates," in Proceedings of the 17th International Conference on Composite Materials, Edinburgh, UK, July 2009.

2.   M. A. Ardakani, A. A. Khatibi, and S. A. Ghazavi, "A study on the manufacturing of glass-fiber-reinforced aluminum laminates and the effect of interfacial adhesive bonding on the impact behavior," in Proceedings of the 11th International Congress and Exhibition on Experimental and Applied Mechanics, pp. 1948–1956, Florida, Fla, USA, June 2008.

3.   T.-W. Shyr and Y.-H. Pan, "Impact resistance and damage characteristics of composite laminates,"Composite Structures, vol. 62, no. 2, pp. 193–203, 2003.

4.   J. E. L. da Silva Junior, S. Paciornik, and J. R. M. d'Almeida, "Evaluation of the effect of the ballistic damaged area on the residual impact strength and tensile stiffness of glass-fabric composite materials,"Composite Structures, vol. 64, no. 1, pp. 123–127, 2004.

5.   L. M. Nunes, S. Paciornik, and J. R. M. d'Almeida, "Evaluation of the damaged area of glass-fiber-reinforced epoxy-matrix composite materials submitted to ballistic impacts," Composites Science and Technology, vol. 64, no. 7-8, pp. 945–954, 2004.

6.   C. Meola and G. M. Carlomagno, "Impact damage in GFRP: new insights with infrared thermography,"Composites A, vol. 41, no. 12, pp. 1839–1847, 2010.

7.   I. M. de Rosa, C. Santulli, F. Sarasini, and M. Valente, "Post-impact damage characterization of hybrid configurations of jute/glass polyester laminates using acoustic emission and IR thermography,"Composites Science and Technology, vol. 69, no. 7-8, pp. 1142–1150, 2009.

8.   L. Krstulovic-Opara, B. Klarin, P. Neves, and Z. Domazet, "Thermal imaging and thermoelastic stress analysis of impact damage of composite materials," Engineering Failure Analysis, vol. 18, no. 2, pp. 713–719, 2011.

9.   S. Wang, D. D. L. Chung, and J. H. Chung, "Impact damage of carbon fiber polymer-matrix composites, studied by electrical resistance measurement," Composites A, vol. 36, pp. 1707–1715, 2005.

10. M. V. Hosur, C. R. L. Murthy, T. S. Ramamurthy, and A. Shet, "Estimation of impact-induced damage in CFRP laminates through ultrasonic imaging," NDT and E International, vol. 31, no. 5, pp. 359–374, 1998.

11. F. Aymerich and S. Meili, "Ultrasonic evaluation of matrix damage in impacted composite laminates,"Composites B, vol. 31, no. 1, pp. 1–6, 2000.

12. Y. Xiong, C. Poon, P. V. Straznicky, and H. Vietinghoff, "A prediction method for the compressive strength of impact damaged composite laminates," Composite Structures, vol. 30, no. 4, pp. 357–367, 1995.

13. A. N. Palazotto, L. N. B. Gummadi, U. K. Vaidya, and E. J. Herup, "Low velocity impact damage characteristics of Z-fiber reinforced sandwich panels—an experimental study," Composite Structures, vol. 43, no. 4, pp. 275–288, 1998.

14. W. A. de Morais, S. N. Monteiro, and J. R. M. d'Almeida, "Evaluation of repeated low energy impact damage in carbon-epoxy composite materials," Composite Structures, vol. 67, no. 3, pp. 307–315, 2005.

15. R. K. Luo, E. R. Green, and C. J. Morrison, "An approach to evaluate the impact damage initiation and propagation in composite plates," Composites B, vol. 32, no. 6, pp. 513–520, 2001.

16. R. K. Luo, "The evaluation of impact damage in a composite plate with a hole," Composites Science and Technology, vol. 60, no. 1, pp. 49–58, 2000.

17. A. MacDowell, X-Ray Micro Tomography, Advanced Light Source, Lawrence Berkley National Laboratory, 2007.

18. G. R. Davis and F. S. L. Wong, "X-ray microtomography of bones and teeth," Physiological Measurement, vol. 17, no. 3, pp. 121–146, 1996.

19. H. Proudhon, "Using X-ray microtomography to probe microstructure and damage of structural materials," in Proceedings of the WEMESURF Contact Course, Paris, France, June 2008.

20. ASTM D7136, "Standard test method for measuring the damage resistance of a fiber-reinforced-polymer matrix composites to a drop-weight impact event," in ASTM Book of Standards, vol. 15.03, 2005.

21. Y. Zhang, A. Johnston, S. Ouellet, K. Williams, D. Boucher, and S. Labonte, "Low-speed impact test for foam supported composite laminates," in Proceedings of the 8th Canada-Japan joint Workshop on Composite Materials, Institute for Aerospace Research, National Research Council Canada, Boucherville, Canada, July 2010.

Chapter 4

Non-Destructive Thermography Analysis of Impact Damage on Large-Scale CFRP Automotive Parts

Alexander Maier[1], Roland Schmidt[2], Beate Oswald-Tranta[2], and Ralf Schledjewski[1, 3]

[1]Chair of Processing of Composites, Montanuniversität Leoben, Otto Gloeckel-Strasse 2, Leoben 8700, Austria

[2]Chair of Automation, Montanuniversität Leoben, Peter-Tunnerstraße 27, Leoben 8700 Austria

[3]Christian Doppler Laboratory for Highly Efficient Composite Processing, Montanuniversität Leoben, Otto Gloeckel-Strasse 2, Leoben 8700, Austria

ABSTRACT

Laminated composites are increasingly used in aeronautics and the wind energy industry, as well as in the automotive industry. In these applications, the construction and processing need to fulfill the highest requirements regarding weight and mechanical properties. Environmental issues, like fuel consumption and CO_2-footprint, set new challenges in producing lightweight parts that meet the highly monitored standards for these branches. In the automotive industry, one main aspect of construction is the impact behavior of structural parts. To verify the quality of parts made from composite materials with little effort, cost and time, non-destructive test methods are increasingly used. A highly recommended non-destructive testing method is thermography analysis. In this work, a prototype for a car's base plate was produced by using vacuum infusion. For research work, testing specimens were produced with the same multi-layer build up as the prototypes. These specimens were charged with defined loads in impact tests to simulate the effect of stone chips. Afterwards, the impacted specimens were investigated with thermography analysis. The research results in that work will help to understand the possible fields of application and the usage of thermography analysis as the first quick and economic failure detection method for automotive parts

INTRODUCTION

Starting its development as an aerospace material, fiber-reinforced and particle-reinforced composites have been increasingly used in other industries, for example automobiles, marine transport, buildings and civil infrastructure, sporting goods, medical equipment and prosthetic devices, etc. With the increased use of composite materials, there is a tremendous need to develop efficient manufacturing techniques, economical and effective repair techniques and methods to predict the short- and long-term behavior of the composite materials and structures made of these materials under a variety of loading and environmental conditions. In addition to the aforementioned requirements, the construction and processing need to fulfill the highest requirements in the weight and mechanical properties [1].

Especially for composite parts in automotive applications, the environmental issues, such as fuel consumption and exhaust emissions, set new challenges in producing lightweight parts that meet the highly monitored standards for these branches. In the automotive industry, one main aspect of construction is the impact behavior of structural parts [2, 3].

Fiber reinforced composites have tremendous strength and stiffness in the plane and main direction of the fibers. However, their strength and stiffness perpendicular to the fibers is governed mainly by matrix properties. Thus, composite structures perform poorly under impact loads that occur due to dropped hand tools, runway debris and hailstone. On the one hand, it is important to understand the threshold impact energy that will cause impact damage and, also, the residual strength after impact damage has occurred, and on the other hand, the detection of impact failure through non-destructive testing methods is necessary. Furthermore, testing the quality of newly designed parts at low cost and time is crucial in competitive fields, such as the automotive field. Therefore, non-destructive test methods are increasingly used [3–5]. As a matter of fact and as confirmed by several independent research works, ultra-sonic testing equipment provides the most information for high quality impact characterization [6–8]. Due to the rising dimension of composite applications, for example rotor plates of wind energy applications or huge structural parts for airplanes and automobiles, it becomes more and more important to find a quick and economic way to detect failures due to impact, which are sometimes barely visible, but nevertheless compromise the composites efficiency. Therefore, non-destructive thermography analysis gives us the ability to do a quick area detection on these parts and, in a further step, using ultrasonic analysis for single-point detection, to gather more detailed information about the amount of damage.

BACKGROUND

Vacuum Infusion

Basic vacuum infusion is an easy method for manufacturing endless fiber reinforced composite parts. Therefore, only a simple, geometry-defining mold, vacuum bag, vacuum pump, resin and reinforcements are needed. In this work, for test specimen production, a slightly modified vacuum infusion was used. The slightly modified concept of vacuum infusion is to spread the resin in large areas by the help of flowing agents and to impregnate the reinforcements through the thickness in a manner more similar to the method description for the

Seaman Composites Resin Infusion Molding Process (SCRIMP). The benefit of this mechanism is that the resin has only a short infusion path through the thickness of the manufactured parts, rather than a long way through the length of the parts. The maximum possible diameters (length, width and height) are dependent on the flow-resistance and the type of resin, as well as the reinforcement structures, the component thickness and the effectiveness of the vacuum [1,9,10].

The basic requirement for vacuum infusion is to guarantee that the mold used and the vacuum infusion structure are as airtight as possible. If this precondition of generating an effective vacuum is not fulfilled, the result will be shown in poorly manufactured parts, e.g., parts that include dry spots and/or entrapped air. Inside that area, the reinforcements are placed in order of the possible variety of loadings that they have to resist. In some cases, to improve the laying down of the enforcement parts, it is possible to use a binder to fix the dry reinforcements on the contour of the mold or to use other preform techniques, for example, like stitching and cutting, net-shape preforming or direct fiber. Normally, with vacuum infusion, composite parts with a fiber percentage up to 60% are possible [11]. As the next step, a resin feed wire and a suction wire have to be placed correctly in the mold to guarantee quick and proper resin infusion and impregnation. As an additional chance to improve the results of the vacuum infusion, simulating the infusion process provides extra information and a possible chance to rethink the whole procedure. On top of this construction, a peel ply, flowing agents, resin absorption material for spare resin and the vacuum bag are the last applied components. After finishing the construction, a vacuum test is necessary to check if the whole mold is airtight, and in the case that the test is positive, completing the impregnation can be started by filling the part with resin by using the suction power of the vacuum. Subsequent to the filling and impregnation process, the resin has to cure, and the manufacturing is done [12]. In the following a schematic drawing, a typical vacuum infusion construction is shown (Figure 1).

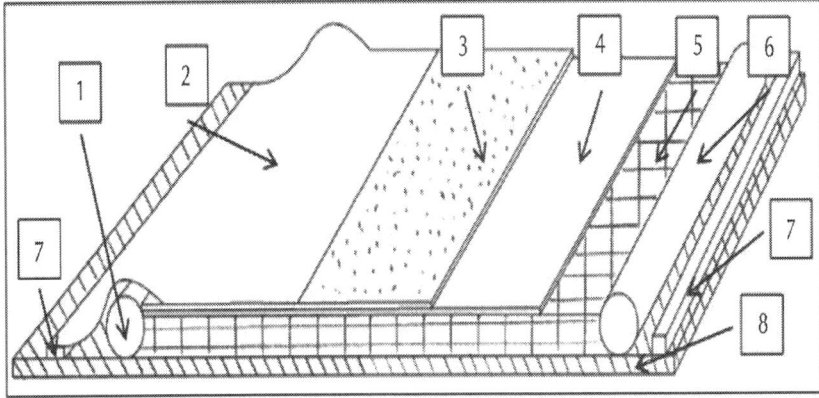

Figure 1: Schematic drawing of a typical vacuum infusion construction; mold assembly: 1, inlet; 2, vacuum bag; 3, distribution medium; 4, peel ply; 5, reinforcement; 6, outlet; 7, vacuum seal; 8, mold die; referring to Correia N.C. et al. (2005) [12].

Impact Test

Impact energies can induce complicated forms of damage, for example, as shown in Figure 2. Damage in composites often begins on the non-impact surface or in the form of an internal delamination, and often, the detection and characterization of impact damages are difficult. The impact damage mechanism in a laminate constitutes a very complex process. It is a combination of matrix cracking, surface buckling, delamination, fiber shear-out and fiber fracture, etc., which usually all interact with each other. The delamination caused by the mismatching of the bending stiffness was propagated and aligned along the direction of the fibers [1,13,14]. The delamination pattern is dependent upon the structure of the fabric. It requires an understanding of the basic mechanics and the damage mechanism [15]. As a matter of fact, composites can fail in a wide variety of modes, because nearly every impact damage mechanism reduces the structural integrity of the component. Even barely visible impact damage can cause fatal destruction to the composite parts, because most composite parts are brittle and, therefore, can only absorb energy in an elastic mechanism or damage mechanism [16]. Impact behavior is even more difficult to understand in applications where hybrid composites are used. In many advanced engineering applications, such as the automotive and aviation industries, hybrid composites have been extensively used due to their high strength, low weight, good fatigue life and corrosion resistance. In addition, their behavior under impact loading has been of significant concern in engineering. There are many studies in the literature [17–22] of the impact response of composite materials and structures. For example, Ying [23] has investigated the damage resistance of three types of carbon laminates and fabric reinforced composites and used finite element analysis coupled with a failure technique to predict the impact threshold energy of the laminates. Low velocity impact damage to composite parts, most of the time, are due to both operational and maintenance activities. In the operational environment, there are typically few incidents of low velocity impact damage, and most can be attributed to hailstone strikes and foreign object damage, such as runway debris, e.g., stone thrown by the tires. The major source of low velocity impact damage for structural composite parts is due to mishandling and maintenance mishaps that include part transportation, handling and storage, and incidental tool drops. However, the impact energy is initially absorbed through elastic deformation till a threshold energy value. Beyond this value, impact energy is absorbed through both elastic deformation and the creation of damage through various failure modes. The type of failure depends on the material and geometric properties of both the impactor and target materials. It has been documented by several investigators that damage initiation is manifested in the load-time history as a sudden load drop due to the loss of stiffness from unstable damage development. Subsequently, damage growth will arrest; the composite laminate will reload, and a cycle of damage propagation and arrest occurs until the impactor begins to rebound and the laminate is unloaded [24,25]. However, damage in composites often begins on the non-impacted surface or in the form of an internal delamination. To gather more information about the damage profile caused by an impact test, many testing procedures are used. Several inspection techniques (acoustic emission, thermography, dye penetrant, stereo X-ray radiography, ultrasonic), with different sensitivity levels, can be used for the non-destructive evaluation of composite materials. All these testing procedures are divided mainly into area detection, e.g., thermography analysis and single-point detection, e.g., ultrasonic analysis. Many investigators have studied the empirical relationship between the initial kinetic energy of the impact projectile and the area of the damaged region detected by mainly single-point detection, like ultrasonic analysis. However, in the case of barely or not visible low impact damages, single-point detection methods are stretching their limits, because screening a large area is not economic and requires too much time. In that case, detecting the damages is part of the area detection, like thermography analysis, and ultrasonic analysis is a part of further analysis to get more information about well-known damage spots [6–8].

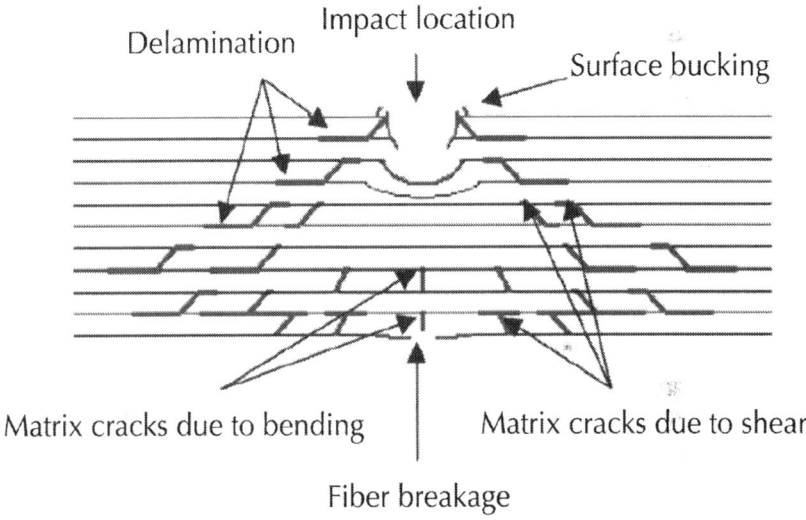

Figure 2: Schematic representation showing a typical impact damage mode for a composite laminate; referring to Tien-Wei S., Yu-Hao P. (2003) [26].

THERMOGRAPHY ANALYSIS

In the case of active thermography inspection, in the first step, heating is applied to the tested specimen, and with an infrared camera, the spatial and temporal distribution of the surface temperature is recorded. As cracks, defects and delaminations disturb the heat flow, evaluating the infrared images can reveal hidden failures. Heating the sample can be carried out optically, e.g., with a short flash pulse, with a laser or with halogen lamps [27]. To electrical conductive materials, such as metals or CFRP (carbon fibre reinforced polymers), heat can be very effectively applied by induced eddy currents. In the case of ultrasound stimulated thermography, cracks are selectively heated by the elastic waves [28]. In many cases, the heat is applied in a short pulse form, but in other cases, it can be sinusoidal modulated to use the so-called lock-in technique [29]. Additionally, in the last few years, different techniques have been developed for how the temporal change of the temperature can be evaluated, in order to obtain high contrast images of subsurface defects [30].

The answer to a short heating pulse, as from a flash lamp, can be evaluated in the frequency domain, with the so-called Pulse Phase Thermography (PPT) [31]. Fourier transformation of the temporal change of the temperature in each pixel of the infrared image sequence defines complex thermal waves. Usually, the phase values of these waves are investigated. As thermal waves with different frequencies penetrate into different depths of the material, phase images belonging to different frequencies exhibit failures in different depths [31].

Another possibility to evaluate in the time domain of the response to a short heating pulse is to calculate first or second derivatives of the temporal temperature change. These make the heat flow and the change of the heat flow visible. However, in order to calculate the derivatives of measured data, as a preliminary step, a polygon fitting is necessary to reduce the noise. This polygon fitting, called Thermographic Signal Reconstruction (TSR) [32], is carried out in the double-logarithmical scale, and it also allows for a very good reconstruction of the images with a low noise level [32].

It has been also shown [27–33] that, first, using a TSR technique and, then, calculating the phase images with Fourier transformation significantly enhances the signal-to-noise ratio of the images and, therefore, increases the detectability of defects deep below the surface.

In our experiments, a flash lamp was used with a 1 ms (full-width half-maximum) heating pulse with an energy of 6 kJ from one xenon flash lamp at a distance of 60 cm from the specimen. The heated side of the CFRP samples was imaged with a commercially available cooled 320 × 256 pixel InSb (indium-antimony detectors) camera, operating at 385 Hz in the 1.5–5 μm spectral range (Figure 3).

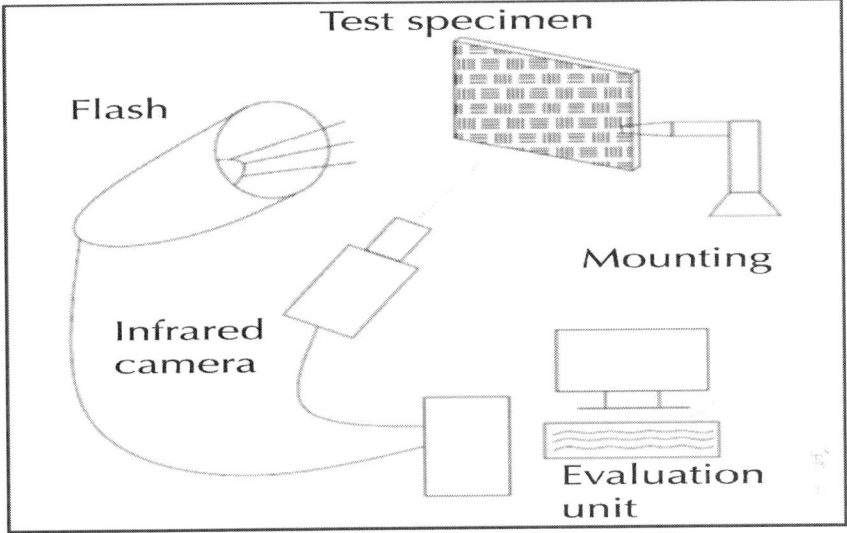

Figure 3: Thermography analysis testing setup.

TEST SPECIMEN PREPARATION

Production of CFRP Testing Specimen

The composite plates were manufactured from 0°/90° twill woven carbon fabrics (style 442 3K Aero) and epoxy resin by the vacuum infusion method. The epoxy resin used was EPIKOTE™ Resin MGS® RIMR 135, and the hardener was EPIKOTE™ Curing Agent MGS® RIMH 135, before the manufacturing process, stored at room temperature. The mixing ratio for resin-to-hardener in weight was 10:3. The composite plates were cured for 24 h at room temperature (23 °C) and under vacuum at a constant−900 mbar pressure.

For the first thermography analysis, the testing specimens were produced with dimensions of 400 mm × 400 mm, 5 layers, and between each layer, small Teflon pieces of 10 mm × 10 mm were positioned to simulate delamination (Figure 4). Due to this stacking, a thickness of the testing specimen of 3.2 mm was reached.

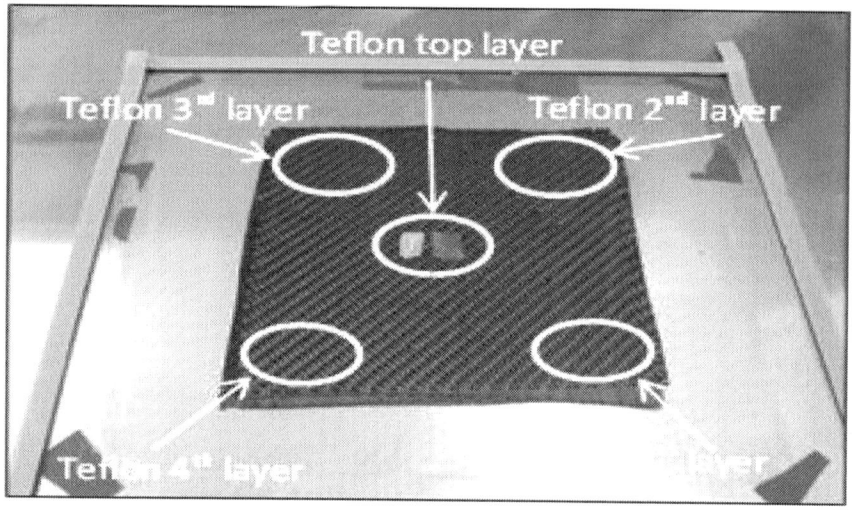

Figure 4: Test specimen, first thermography analysis.

For the second part of this work, stacking sequences for the test specimen for the impact tests of (C_0/C_{90})-woven at $45°$ + (C_0/C_{90})-woven at $0°$ and, again, (C_0/C_{90})-woven fabrics at $45°$ were chosen. This layer configuration of the composite plates was given due to the operating condition of a real automotive part produced in the Cars Ultra-light Technology project (CULT). After the manufacturing process, the composite specimens with dimensions of 150 mm × 100 mm were trimmed from the laminated plates.

IMPACT TESTING

In this work, all low velocity impact tests were based on the ÖNORM EN ISO 6603:2 [34,35], and an Instron-CEAST 9350 impact testing machine (Instron GmbH, Pfungstadt, Germany) was used for impact testing. This testing machine consists of a dropping crosshead with its accessories, a pneumatic clamping fixture, a pneumatic rebound brake and an impulse data acquisition system. The weight of the crosshead is adjustable with the drop mass, and the tup of the impactor has a 20.0 mm diameter hemispherical nose. The self-identifying load-cell capacity is 15.56 kN, and the total mass of the impactor with its accessories was kept constant at 2.045 kg for all tests. The test machine has a pneumatic rebound brake system to prevent repeated impacts on specimens. CEAST View 5.94 3C is a software program that records the electronic signals (load vs. time data and instantaneous velocity at the moment of impact). The software program, based on Newton's second law and kinematics, is used to convert the load-time data into a load-deflection relation, with the assumption that the impactor is rigid. The electronic signals are used by the software to calculate the deflection, tup velocity and the energy absorbed by the specimen. For the conducted impact tests, the impact energies varied from approximately 1 J to 5 J or even up to the complete perforation of the specimens. Therefore, it becomes possible to examine the damage mechanisms of composite plates under various impact energies.

As shown in Figure 5, it was not possible to detect matrix cracking, surface buckling, fiber shear-out and fiber fracture from the impact results for the impact loads of 1 J to 2 J. The test series under a 1 J and 2 J impact load showed no typical impact response. As a reason for these responses, we found out that the resolution of the testing machine was not great enough for measuring the impact at such low impact energies. Therefore, for the impact tests of 1 J and 2 J, no further evaluations were possible, and the testing series were not used for further analysis. For the impact results for the impact load of 3 J, it was not possible to detect matrix cracking, surface buckling, fiber shear-out and fiber fracture either, but the testing resolution was high enough, and therefore, we decided to use the impact results for further analysis. For all tests at 4 J and 5 J, at least one aforementioned mechanics failure was detected. For all of the impact tests from 3 J to 5 J, delamination between any of the layers was possible, and therefore, the thermography analysis came in handy. The capabilities of this non-destructive testing method are shown in this work.

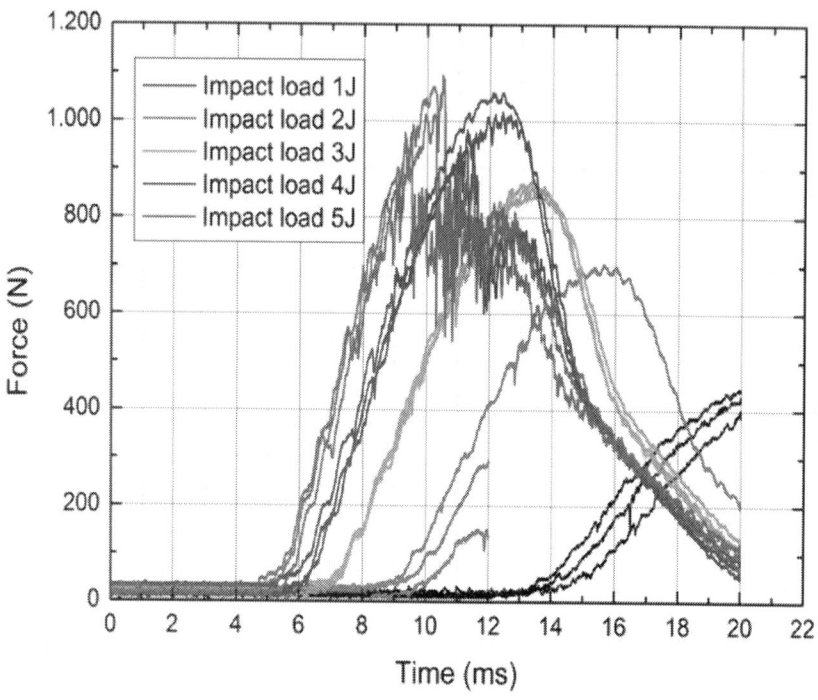

Figure 5: Impact result on the carbon test specimen for 1 J to 5 J.

RESULTS OF THE THERMOGRAPHY ANALYSIS OF TEFLON TESTS

The advantages of the thermography testing are that not only the subsurface defects can be made visible, but also, according to the delay of the response, their depths can be estimated. Teflon inserts are often used in CFRP structures to create artificial delaminations. A specimen, as shown in Figure 3, has been prepared with five layers and with Teflon inserts between them. The temperature response to a flash pulse heating has been evaluated by the PPT technique, following a TSR step, to increase the signal-to-noise ratio of the images. Figure 6 shows the results, and the figures from left to right demonstrate that with the evaluation of different frequencies, it is possible to make deeper and deeper images of the delaminations.

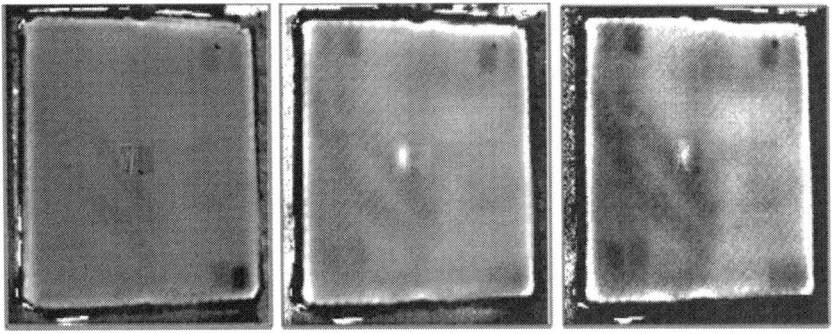

Figure 6: Phase images, after using Thermographic Signal Reconstruction (TSR), of the sample with Teflon inserts, shown in Figure 4. The frequencies of the images are 0.1 Hz (left), 0.15 Hz (middle) and 0.18 Hz (right).

RESULTS AND DISCUSSION THERMOGRAPHY ANALYSIS/IMPACT TESTS

Load/Time Impact Result (F/t–Diagram)

Figure 7 shows the Load-time (F/t) curves of the carbon testing specimens for thermography analysis at different levels of impact energy as an extract of Figure 5, because in this study, the results will be based on these impact energies. Individually, each curve has an ascending section of loading, reached a maximum load value and has a descending section of unloading. The ascending section of the Load/time curve is called the bending stiffness, due to the resistance of the composite to impact loading, and at this section, the maximum load value reached the highest maximum load; this value is called the peak force.

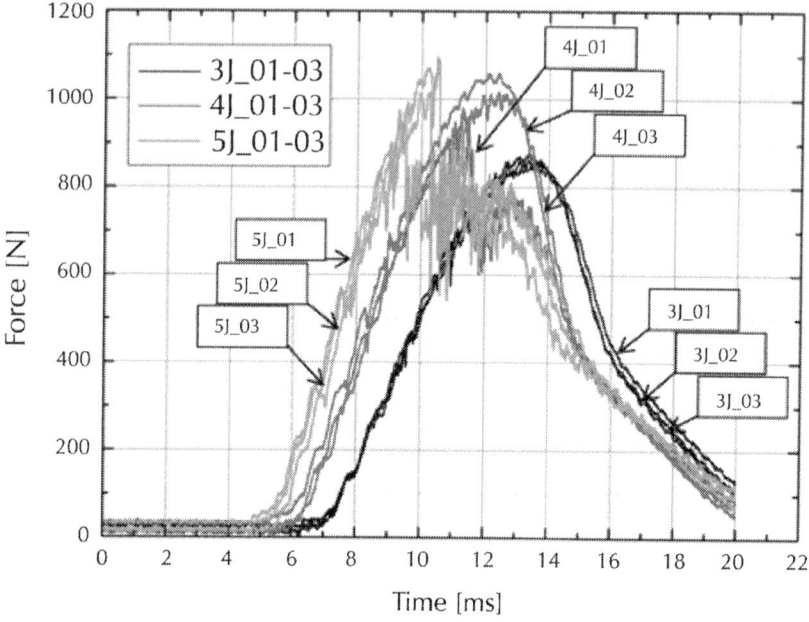

Figure 7: Load/time (F/t) curves of the carbon specimen for thermography analysis.

Typical Load/time curves of composite plates subjected to impact loading have three situations, including rebounding, penetration and perforation. Furthermore, load-time curves, in general, can be classified as closed-type curve and open-type curve. The rebounding case results in closed curves, indicating the rebounding of the impactor from the specimen surface. The closed-type curves return back from the maximum load or the peak force value towards the axis of abscissas means to 0 N force without a sudden load drop due to the loss of stiffness from unstable damage development. These curves are all curves with 3 J impact energy (3J_01–03), and, at an impact energy of 4 J, the second and the third one (4J_02, 03) are as seen in Figure 7.

When the impact energy increases, closed-type curves bound larger areas, and deflection increases while the rebounding section becomes smaller. As seen from Figure 7, at an impact energy of 4 J, there are closed-type (4J_02, 03) curves, but also open-type curves (4J_01). From that point on, as the impact energy continuously increases, the curve type changes from the closed-type to the open one. If a curve is of an open type, the specimen is either penetrated or perforated by the impactor. Therefore, the testing specimen at 5 J (5J_01–03) represents the penetrated case, while the others represent the perforated case, as seen in Figure 7.

Optical Impact Characterization and Thermography Analysis

The damage profiles of damaged composite specimens were evaluated from the front (impacted) and back (non-impacted) side by visual inspection and thermography analysis. In general, impact damage modes consist of indentation, matrix cracking, delamination between layers, fiber pull-out and fiber breakages. Impact-induced damage, which may be undetectable by visual inspection, can have a significant effect on the strength, durability and stability of the composite structure. Again, due to the fact that this work is about the detection of failure in general, there is no further examination of which type of failure is detected after the detection itself. In the following paragraphs, for understanding the possibilities and importance of non-destructive thermography analysis with respect to the possibility of doing quick and keen area failure detection, several images of damaged specimens were compared and explained with the load-time curves, visual detection and thermography analysis results. Therefore, the carbon testing specimens at impact energies of 3 J, 4 J and 5 J were used.

At the highest impact energy during this work (5J_01-03), penetration and perforation started to take place for the carbon testing specimen, as shown in Figure 7. The penetration threshold can be defined as the energy level when the impactor gets stuck in the specimen for the first time and does not rebound from the specimen surface any more. The perforation threshold can be defined as the energy level at which the impactor passes through the thickness of the specimen for the first time resulting in permanent catastrophic damage to the specimen. As seen in the visual inspection (Figure 8), all carbon fibers are damaged through the thickness, but compared to the F/t curves, the energy was not enough for the impactor to get stuck in the specimen; a short rebound is seen in Figure 7 for the 5 J tests. Due to the visual inspection, the perforation process started at this point, but a complete perforation (catastrophic damage) was not seen. These results from the impact test and the visual detection are congruent with the result given by the thermography analysis, as seen in Figure 9. As additional information, the bottom pictures in Figure 9 showed two horizontal line marks on the specimen, which refer to the placement of the flowing agent during the vacuum manufacturing. These marks came from the vacuum under pressure and are due to the compaction of the rims of the flowing agent being marked on the composite plates. That result is not a part of this study and, therefore, not evaluated further. The same applies to Figures 11 and 12. However, the thermography analysis displays the damage section (fiber breakage, penetration, lamination, etc.) perfectly as a cross in the center of the impact zone. The four pictures at different times show the damage profile through the specimen thickness and gives a good overview that the impact damage goes through the whole testing piece, and as a conclusion, the thermography analysis is comparable with the F/t curves from the impact tests, therefore, for this stage of damage, the possibility for non-destructive damage detection.

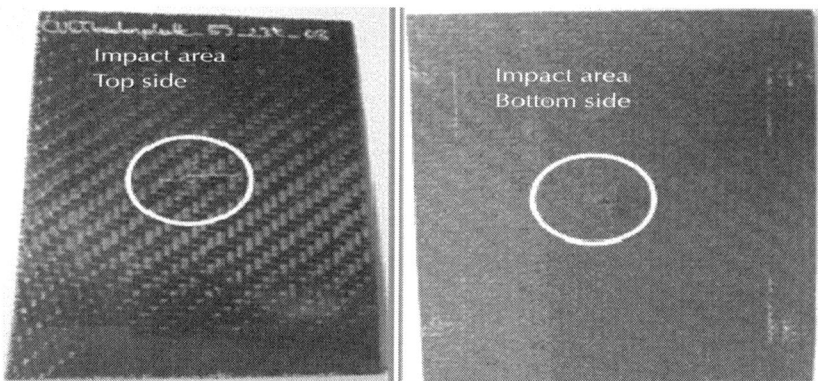

Figure 8: Visual failure detection for an impact energy of 5 J; (left) top side up; and (right) bottom side up.

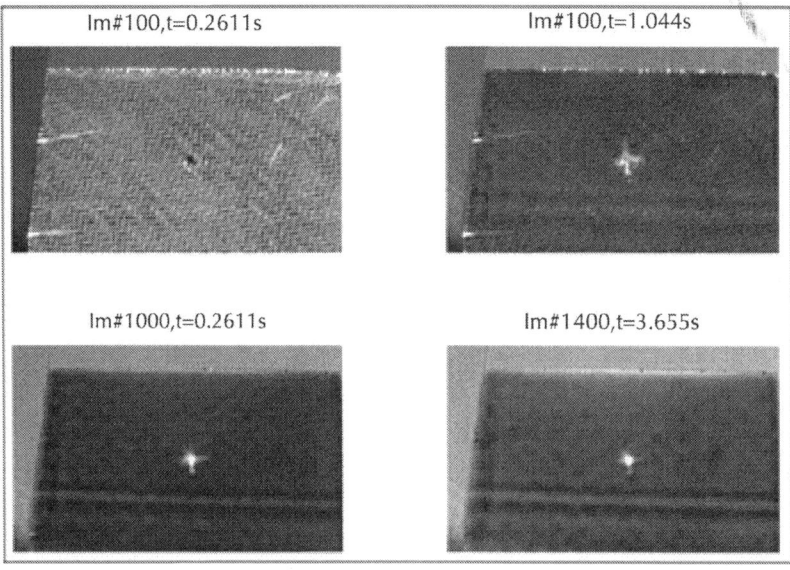

Figure 9: Catastrophic damage of the specimen shown by the thermography analysis after impact testing at 5 J, the second derivative of TSR.

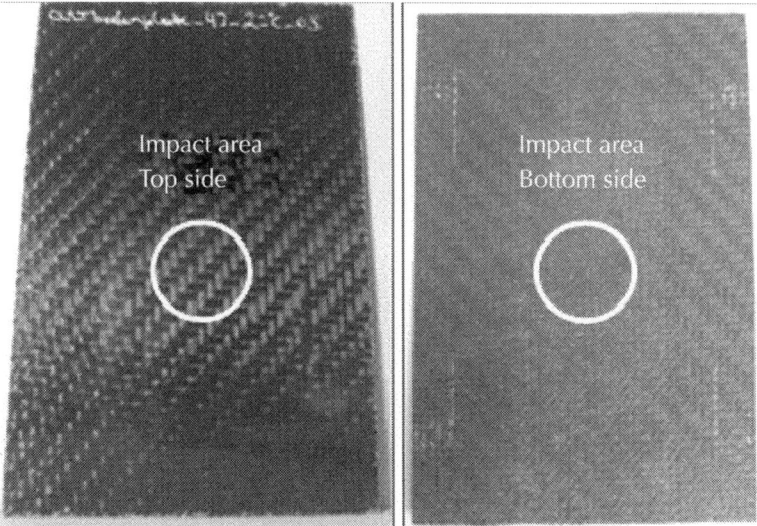

Figure 10: Visual failure detection for an impact energy of 4 J; (left) top side up; (right) bottom side up.

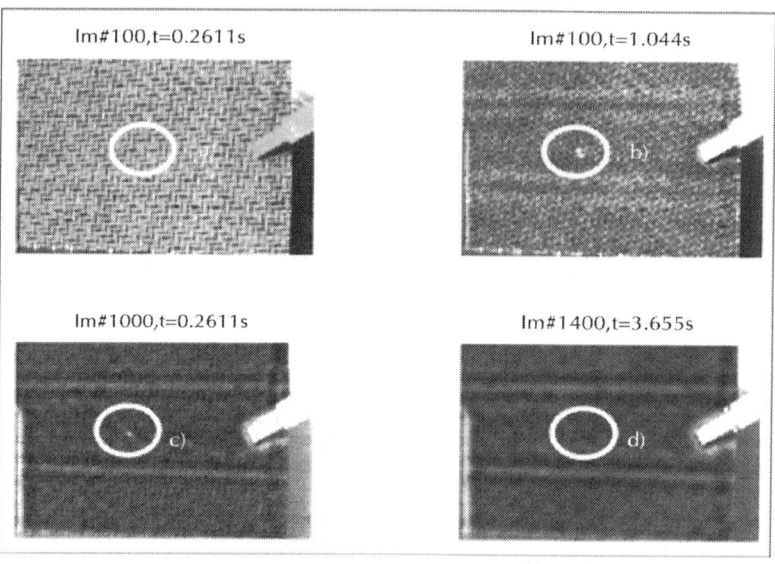

Figure 11: Thermography analysis after impact testing at 4 J, the second derivative of TSR.

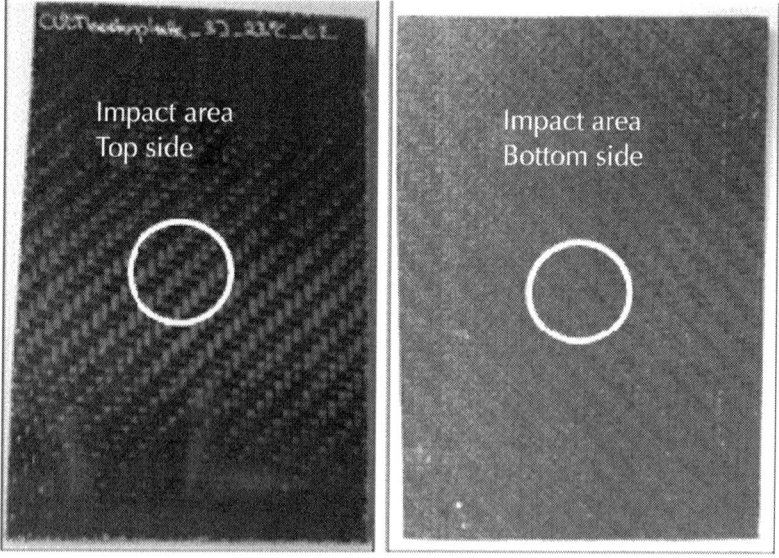

Figure 12: Visual failure detection for an impact energy of 3 J; (left) top side up; and (right) bottom side up.

As the impact energy decreases (4 J), the F/t curves of composites shrink in the negative direction of the horizontal axis, and due to the fact that the damaged composites had to absorb less impact energy, the damage profile of the catastrophic damage picture does not immediately show failure, as expected. F/t curves for 4 J compared with the visual inspection show the same tendency of lesser damage at lower impact energies, except for the testing specimen, 4J_03. These impact tests showed an open F/t curve, and therefore, the beginning of penetration and perforation would be expected. In Figure 10, as seen in the sectioning of the damage, there are no fiber breakages or penetration through the thickness of the carbon layers. The visual inspection of the other 4 J test specimen (4J_01, 02) showed the same result as the specimen, 4J_03. There are only small scratches, and on closer examination, matrix cracking and impact signs (small dots) of the impactor were shown, as seen in Figure 10. Regarding the F/t curves, it seemed that around an impact energy of 4 J, the damage profile is in a transition area between catastrophic damage and possible hidden damage inside the carbon layers. Therefore, heavy damage inside the specimen is possible, but with the F/t curves and the visual inspection, not verified. In part (a) of Figure 11, the thermography analysis showed no damage sign, and due to that very first image, it seemed like the same result as the F/t curves and the visual inspection. Analysis through the thickness of the test specimen thermography pictures (b) and (c) (more time leads to the inner carbon layer) showed damage, seen in Figure 11, as a highlighted spot. That spot means that between the layers, there is delamination caused by the impact test before. In that case, the non-destructive thermography analysis gave the possibility of detecting existing damages inside the specimen, which were not detectable with conventional testing methods. Going on to the last part (d), the thermography did not detect damage at all, and as a conclusion, and compared to the visual detection, the test specimen had no damaged top or bottom layer; only delamination between layers occurred.

For lower impact energies (less than approximately 3 J), the primary damage mode was indentation-induced matrix cracking on impacted surfaces. There were minor matrix cracks on the bottom side of the test specimen, and some delamination could be possible between the inner layers (Figure 12). As shown in Figure 7 (3J_01–03), the impact curve is a closed one, and compared with the visual detection of the top side up or the bottom side up (Figure 12), there is no penetration and/or perforation. As a result of these analyses, there are no failures detected. Nevertheless, interior layer delamination could be possible, but was not detected with the mentioned methods.

At that point, the degree of resolution of the thermography analysis provides, once again, further information of the failure profile of this testing specimen, shown in Figure 13. With the thermography analysis, a small failure crack at the bottom side of specimen was found. In fact, that there is no sudden drop in the F/t curve from the impact test right after peak force, which would detect fiber breakage in interior layers or perforation/penetration, none of these failures were expected. Referring to the visual inspection, the little black dot shown in the right pictures of Figure 13, there is no matrix cracking. One possibility was that this failure could be a delamination between the layers. In Figure 13, the right side, the different parts (a–d) show images along time, and therefore, it could be possible to provide information about the horizontal failure spread. However, for this impact energy, separated from the other analysis results, the non-destructive thermography analysis showed that there is a failure mark due to the impact. If this specimen were a real carbon composite part in a car, as the next step, the impact area, especially, and the area around it should be further inspected, and based on that, a statement about whether this part could be used further on could be made.

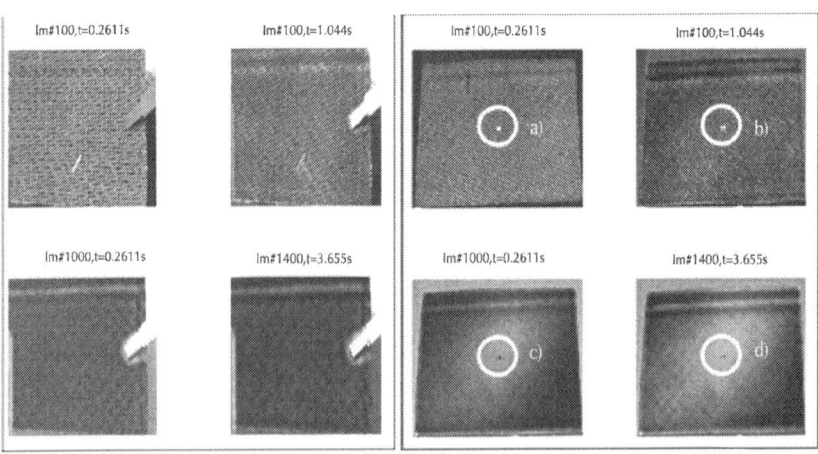

Figure 13: Thermography analysis after impact testing at 3 J; (left) top side up; and (right) bottom side up; the second derivative of TSR.

CONCLUSIONS

This experimental study deals with the investigation of impact damage composite plates using non-destructive thermography analysis to show the ability of this detection mechanism to detect low impact damages in a quick and keen way by using area detection. The purpose of this work was to find these damaged spots and provide position information for further damage analysis, which may gather more information about the damage profile than thermography analysis does. This could be, for example, an ultrasonic-based detection mechanism, the main disadvantage of which is the lack of a quick possibility to spot barely or no visible damages in large areas, because of the single-point detection sensors. The following conclusions can be made from the tests:

- For lower impact energies (up to 3 J), impact events were elastic, and excessive impact energy was used for the rebounding of the impactor. As the damaged specimens were visually inspected, maybe, minor matrix cracking or impactor signs were observed on the carbon surface of the impacted side of the specimens. However, the thermography analysis showed, in a quick way, that there was delamination between the inner layers of the carbon specimen. Due to that fact, further examinations on that part should be done.

- At increasing impact energies (approximately 4 J), the damage profile is in a transition, where damage detection with visual inspection was possible, and the F/t impact curves gave further information about possible damage. In that case, thermography analysis can give more detailed information, but direct use of ultrasonic analysis would be more efficient, because the damages were already spotted by the visual inspection.

- Impact energies around 5 J and above will cause catastrophic damage to the testing specimen, and for inspection, the visual methods were the fastest way and accurate enough.

- The non-destructive thermography testing method showed its potential, especially at lower impact energies because it is a quick, keen and non-destructive area detection method. A possible field of operation is the inspection of large carbon parts in automotive and/or airplane applications.

The authors would like to thank the Institute of Material Science and Testing of Polymers for supporting this study with knowledge and impact testing equipment.

REFERENCES

1. AVK, Industrievereinigung Verstärkte Kunststoffe, Handbuch Faserverbundkunststoffe (in German Language); GWV Fachverlag GmbH: Wiesbaden, Germany.

2. Sankar, B.V. Low-velocity impact response and damage in composite materials. Key Eng. Mater 1996.

3. Sankar, B.V.; Hu, S. Dynamic delamination propagation in composite beams. J. Compos. Mater 1991, 25, 1414–1426.

4. Sankar, B.V.; Sonik, V. Pointwise energy release rate in delaminated plates. AIAA J. 1995, 33, 1312–1318.

5. Maldague, X. Theory and Practice of Infrared Technology for Non-Destructive Testing; John Wiley & Sons, Inc: Hoboken, NJ, USA, 2001.

6. Aymerich, F.; Meili, S. Ultrasonic evaluation of matrix damage in impacted composite laminates. Compos. B Eng 2000, 31, 1–6.

7. Michel, C.; Singh, D.; Viot, P. Sizing of impact damages in composite materials using ultrasonic guided waves. NDT & E Int 2012(46), 22–31.

8. Catherine, P.; Chotard, T.; de Belleval, J.-F.; Benzeggagh, M. Characterization of composite materials by ultrasonic methods: modelization and application to impact damage. Compos. B Eng 1998, 29, 159–169.

9. Bhattacharyya, D.; Fakirov, S. Synthetic Polymer-Polymer Composites; Carl Hanser Verlag: München, Germany, 2012.

10. Mitschang, P.; Neitzel, M. Handbuch Verbundwerkstoffe; Carl Hanser Verlag: München, Germany, 2004.

11. Breuer, U.; Neitzel, M. Die Verarbeitungstechnik der Faser-Kunststoff-Verbunde; Carl Hanser Verlag: München, Germany, 1997.

12. Correia, N.C.; Robitaille, F.; Long, A.C.; Rudd, C.D.; Simacek, P.; Advani, S.G. Analysis of the vacuum infusion moulding process:

I. Analytical formulation. Compos. A Appl. Sci. Manuf 2005, 36, 1645–1656.

13. Lesser, A.J. Effect of resin crosslink density on the impact damage resistance of laminated composites. Polym. Compos 1997, 18, 16–27.

14. Liu, D. Impact-induced delamination—A view of bending stiffness mismatching. J. Compos. Mater 1998, 22, 674–691.

15. Maio, L.; Monaco, E.; Ricci, F.; Lecce, L. Simulation of low velocity impact on composite laminates with progressive failure analysis. Compos. Struct 2013, 103, 75–85.

16. Richardson, M.O.W.; Wisheart, M.J. Review of low-velocity impact properties of composite materials. Compos. A Appl. Sci. Manuf 1996, 27, 1123–1131.

17. Abrate, S. Impact on Composite Structures; Cambridge University Press: Cambridge, UK, 1998.

18. Larsson, F. Damage tolerance of a stitched carbon/epoxy laminate. Compos. A Appl. Sci. Manuf. 1997, 28A, 923–934.

19. Liu, D.; Raju, B.B. Effects of joining techniques on impact perforation resistance of assembled composite plates. ExpMech 2000, 40, 46–53.

20. Liu, D.; Raju, B.B.; Dang, X. Size effects on impact response of composite laminates. Int. J. Impact Eng 1998, 21, 837–854.

21. Mili, F.; Necib, B. Impact behaviour of cross-ply laminated composite plates under low velocities. Compos. Struct 2001, 51, 237–244.

22. Reid, S.R.; Zhou, G. Impact Behaviour of Fibre-Reinforced Composite Materials and Structures; CRC Press (Chemical Rubber Company): Boca Raton, FL, USA, 2000.

23. Ying, Y. Analysis of impact threshold energy for carbon fibre and fabric reinforced composites. J. Reinf. Plast. Compos 1998, 17, 1056–1075.

24. Schoeppner, G.A.; Abrate, S. Delamination threshold loads for low velocity impact on composite laminates. Compos. A Appl. Sci. Manuf 2000, 31, 903–915.

25. David-West, O.S.; Nash, D.H.; Banks, W.M. An experimental study of damage accumulation in balanced CFRP laminates due to repeated impact. Compos. Struct 2008, 83, 247–258.

26. Tien-Wei, S.; Yu-Hao, P. Impact resistance and damage characteristics of composite laminates. Compos. Struct 2003, 62, 193–203.

27. Ibarra-Castanedo, C.; Piau, J.-M.; Guibert, S.; Avdelidis, N.P.; Genest, M.; Bendada, A.; Maldague, X.P.V. Comparative study of active thermography techniques for the nondestructive evaluation of honeycomb structures. Res. Nondestruct. Eval 2009, 20, 1–31.

28. Mignogna, R.; Green, R.; Henneke, E.; Reifsnider, K. Thermographic investigations of high-power ultrasonic heating in materials. Ultrasonics 1981, 7(1), 159–163.

29. Wu, D.; Busse, G. Lock-in thermography for nondestructive evaluation of materials. Revue Générale Thermique 1998, 37, 693–703.

30. Vavilov, V.P.; Almond, D.P.; Busse, G.; Grinzato, E.; Krapez, J.-C.; Maldague, X.; Marinetti, S.; Peng, W.; Shirayev, V.; Wu, D. Infrared Thermographic Detection and Characterisation of Impact Damage in Carbon Fibre Composites: Results of the Round Robin Test, In Proceedings of the 4th Conference on Quantitative Infrared Thermography (QIRT), Lodz, Poland, 7–10 September 1998.

31. Maldague, X.; Marinetti, S. Pulse phase infrared thermography. J. Appl. Phys 1996.

32. Shepard, S.M. Temporal Noise Reduction, Compression and Analysis of Thermographic Image Data Sequences. US Patent 6,516,084, 4 February, 2003.

33. Oswald-Tranta, B.; Shepard, S.M. Comparison of Pulse Phase and Thermographic Signal Reconstruction Processing Methods. SPIE Proceed 2013, 8705.

34. ÖNORM EN ISO 6603-2:2002 04 01, Kunststoffe—Bestimmung des Durchstoßverhaltens von festen Kunststoffen—Teil 2: Instrumentierter Schlagversuch (ISO 6603-2:2000) (in German Language); Austrian Standards plus GmbH: Vienna, Austria, 2002.

35. ISO 6603-2:2000, Plastics—Determination of Puncture Impact Behaviour of Rigid Plastics—Part 2: Instrumented Puncture Test; International Organization for Standardization: Geneva, Switzerland, 2000.

5

Optical Fiber Embedded in Epoxy Glass Unidirectional Fiber Composite System

Irina Severin[1, 2], Rochdi El Abdi[2], Guillaume Corvec[2], and Mihai Caramihai[1]

[1]Department Materials Technology/Computer Science, Politehnica University of Bucharest, 313 Splaiul Independentei, Bucharest 060042, Romania

[2]Larmaur, ERL CNRS 6274, University of Rennes 1, Beaulieu Campus, Rennes Cedex 35042, France

ABSTRACT

We aimed to embed silica optical fibers in composites (epoxy vinyl ester matrix reinforced with E-glass unidirectional fibers in mass fraction of 60%) in order to further monitor the robustness of civil engineering structures (such as bridges). A simple system was implemented using two different silica optical fibers (F1—double coating of 172 µm diameter and F2—single coating of 101.8 µm diameter respectively). The optical fibers were dynamically tensile tested and Weibull plots were traced. Interfacial adhesion stress was determined using pull-out test and stress values were correlated to fracture mechanisms based on SEM observations. In the case of the optical fiber (OF) (F1)/resin system and OF (F1)/composite system, poor adhesion was reported that may be correlated to interface fracture at silica core level. Relevant applicable results were determined for OF (F2)/composite system.

INTRODUCTION

Composite materials are now becoming accepted for use in many major structural applications, particularly in the space, aerospace, marine, civil engineering, and automotive industries. These materials with their high specific strength and stiffness are gradually replacing their metal counterparts due to reduced weight, improved wear and corrosion resistance, increased fatigue life, and the ability of the material and component to be formed at the same time. The wide ranges of structural composites are based on polymeric matrices with glass or carbon reinforcement [1].

Optical fiber sensors have gained much attention in recent years for a variety of physical and chemical measurements. Intense research has been performed to develop "special fibers" for integration in smart structures in order to replace conventional sensors [2–4] and to provide information on the continuous evolution of damaged structures under mechanical stress, thus monitoring structural robustness [5,6].

The efficiency of optical fiber sensors to continuously monitor structure deformation and reinforcement cracks depends mainly on strength transfer at the interface level between the embedded optical fiber and the composite structure [7,8]. In order to assess mechanical properties at interface level and composites performance, certain testing, such as the pull-out test, has been developed [9,10]. The test may be used to reveal the bonding quality between the optical fiber sensor and the composite structure. The interfacial adhesion strength τ_d appears as a critical factor for structural robustness monitoring being given by:

$$\tau_d = F_d/2 \cdot \pi \cdot R \cdot L_e$$

(1)

where F_d is the maximum force corresponding to the linear zone of the force *versus* displacement curve; R is the core optical fiber radius and L_e the optical fiber embedded length.

In the context of recently developed optical fibers for distributed sensors and other smart structures, the reliability of optical fibers appears as a critical factor. Testing procedures and damage mechanisms identified on a series of experimental studies are given elsewhere [11].

The present study has the aim to manage a reliable fabrication process of embedding optical fiber in an epoxy/glass fiber composite and to obtain basic data on a simple system. In a further step, Bragg grating optical fiber is envisaged to be embedded in a similar composite structure in order to implement a smart structure sensitive to flexion stress.

EXPERIMENTAL PROCEDURE

Composite Fabrication

The polymer matrix used in this study, provided by the French company DFC-DCP Pultrusion (Creil, France), is a mixture of epoxy vinylester resin DERAKANE 470-36 and a catalytic system composed of Styrene, Perkadox 16 and Trigonox C. After homogenously stirring the epoxy resin, the powdered Perkadox 16 is firstly diluted in Styrene so as to form a homogeneous Styrene/Perkadox 16 system, and then the Trigonox C is added. For the experimental implementation, small matrix quantities were prepared, such as 30 g of epoxy resin and 0.66 g Styrene, 0.264 g Perkadox 16 and 0.108 g Trigonox C.

Composite samples were prepared, reinforcing the epoxy matrix with E-glass fibers Roving 4800 Tex having a density of 2.62 g/cm^3. A mass fraction of 63% glass fiber in epoxy matrix was envisaged, meaning an incorporation of 12 roving in the implemented composite sample. To calculate the roving number, the following relation is used:

$$N = 0.545833 \, f \, V_{com}$$

(2)

where f is glass fiber volume fraction and V_{com} the composite volume. Roving 4800 Tex means 1 roving = 0.0048 g/mm. For the samples of 50 mm × 5 mm × 10 mm, 12 roving correspond to a volume fraction of 40% and a mass fraction of 63%, respectively. The composite samples were weighted after fabrication and effective mass fraction ranging between 55% and 65% were obtained.

In order to characterize the interfacial adhesion between the optical fiber and the epoxy vinylester/glass fiber composite material (OF/C), a simple system composed of a single optical fiber embedded in epoxy vinylester resin (OF/R) was preliminarily studied. The system is a unidirectional composite, where glass fibers have the sample length direction and the optical fiber is centrally embedded in the reinforcement direction.

Two different optical fibers elaborated by iX Fibers S.A.S. company (Lannion, France) were used: firstly, the fiber, denoted by F1, having a 80 μm silica core and two acrylate polymer layers of 172 μm secondary coating diameter and secondly, the fiber, denoted by F2, having a 80 μm silica core, but a single acrylate polymer layer of 101.8 μm fiber coating diameter.

As optical fiber surface has determined fracture to a large extent, external coating appears critical. This coating is polymeric in most cases, and modern optical fibers are coated by two different layers: a soft coating at glass surface and a hard coating at external surface. The coating makes a protection against scratches that occur in normal handling and it reduces water activity at glass surfaces. Polymeric coatings (e.g., epoxy-acrylate) are preferred in practice because they are more efficient at inhibiting surface defects. As one will notice later in our study, regarding MEB examinations (see Section 3.3), fiber F1 has the typical section with two layers, but fiber F2 was subjected to a thermal treatment and presents one layer surrounding the silica core.

The thermal treatment consisted of exposure at 200 °C curing temperature for 100–200 h. The fiber behavior appears similar to polyimide, those polymers containing chemical groups that promote solubility without affecting significantly mechanical properties.

Moreover, curing at 200 °C for 100 h, leads to the lost of 1.5% of the mechanical strength as compared to 9% lost in the case of polyimide fiber. Fiber parameters are given in Table 1.

Table 1: iX Fiber reference

Parameter	F1	F2
Optical fiber reference	iXF-PMG-1550-80-V20	iXF-PMG-1550-80-HT
Attenuation@1550 nm, dB/km	1.15	1.68
Cutoff wavelength, nm	1300	1318
Bare fiber/coating diameter, µm	80/172	80/101.8

The sample preparation consisted of incorporating the glass fiber in the resin into a mold and embedding the optical fiber centrally positioned into the sample. As seen in Figure 1, the glass roving is placed in the mold and, superimposed layer by layer (6 roving), is carefully impregnated with the resin until half the mold is filled, and then a fine steel cylinder guide is placed on top in order to allow optical fiber insertion. The mold is filled in, as previously described (other 6 roving), and then the optical fiber is gently inserted in the fine guide. Before closing the mold with a controlled screw torque of 2 Nm, the steel guide is gently and carefully extracted. For safety reasons, all operations were performed into an extractor hood.

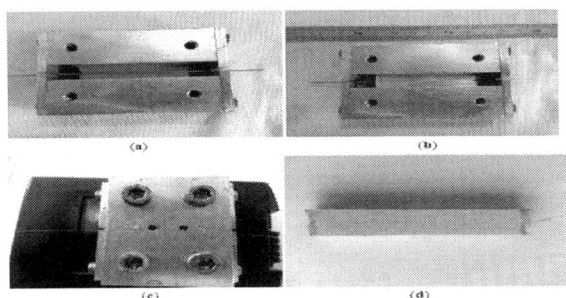

Figure 1: Composite manufacturing experimental procedure. (a) Pre-impregnated E-glass roving; and optical fiber steel guide placement in the mold (b)

mold fill-in; (c) mold closure—see optical fiber ends and resin surplus evacuation system and (d) composite sample.

Then, curing in preheated oven at 80 °C temperature for 90 min was applied and samples were extracted after 60–90 min.

Pull-Out Testing

Following the procedure, series of F1 optical fiber embedded in resin and in composite, respectively and F2 optical fiber embedded in composite were prepared in order to perform pull-out tests. Using a LLOYD LR 50K tensile testing setup (Lloyd Instruments Ltd., Fareham, Hampshire, UK), a gripping force was applied, as seen in Figure 2a. The optical fiber end is rolled up on a pulley of 50 mm in diameter and covered with a powerful adhesive so as to prevent fiber slip during testing. When the optical fiber was pulled-out, the fiber breakage occurred at the bit. Thus, a notch was introduced into the specimen as seen inFigure 2b, at a distance of 8 mm from the sample end. The cross head speed was set at 1 mm·min^{-1}, which corresponds to a strain rate of about 0.02 min^{-1}. The de-bonding force F was considered as the maximum force preceding partial de-bonding. A scanning electron microscope (SEM) (Field Emission Scanning Electron Microscope-type JEOL JSM 6301F, Tokyo, Japan) was used to investigate the surface of optical fibers after pull-out tests.

(a) (b)

Figure 2: (a) Pull-out testing; and (b) sample prepared for pull-out tests.

Optical Fiber Dynamic Tensile Testing

A tensile bench Lloyds Instruments LR 50K (maximum: 100 N) was used. In order to be dynamically tensile tested, sample fibers were three tours rolled on the cylinder pulley having 50 mm in diameter. The pulleys were covered with a powerful double face adhesive; the mechanical properties of the adhesive layer appeared to be important controlling factors. Despite the standard conditions, for economy and testing time reasons, sample testing free length was chosen to 200 mm. A testing strain rate of 20, 100, 200, respectively 500 mm/min was chosen. These rates were selected in order to correspond to 10, 50, 100, respectively 250%/min as compared to sample main length. In the case of the reference fiber, at least 20 samples were tested and Weibull plots were traced, then n_d-stress (n_d, dynamic stress corrosion parameter) corrosion factor was calculated.

In order to briefly evidence several parameter influences (oven treatment for polymerization or catalytic system—Trigonox C) on optical fiber mechanical properties, series of F1 fiber were treated, then tensile tested and compared to as-received fiber.

Results treatment is then usually based on Weibull plots, despite some opponents concerning the adequacy of Weibull distribution use in case of tensile experiments [12–14]. Even if at least 50 samples are proposed for a reasonable estimation using Weibull, due to our mainly aim of comparing fiber mechanical strength subjected to different etching environments, the 20 series used for Weibull treatment appeared relevant.

EXPERIMENTAL RESULTS AND DISCUSSION

Optical Fiber Characterization

Based on our previous experimental studies, the optical fiber was dynamically tensile tested for four different strain rates of 0.1, 0.5, 1 and 2.5 min^{-1} corresponding respectively to the tensile testing cross speed of 20, 100, 200 and 500 mm/min. For each tested sample, we determined the stress to fracture, and then the results were treated through a statistical approach using the Weibull theory. The classical Weibull plots of the logarithm function of the cumulative failure probability (F_k) related to the logarithm of the stress to fracture (C) has allowed us to calculate and to compare the statistical parameters and the n_d-stress corrosion parameter.

It is assumed that fracture at the most critical flaw on a fiber leads to total failure. For brittle materials as silica optical fibers, strength results obtained from tensile tests present a significant scattering. Then, the statistical Weibull method is commonly used. This method leads us to obtain the mean stress value (strength at 50% fracture probability of the Weibull plot), the medium stress, the Weibull slope and the distribution of the critical flaw size in the sample. The statistical Weibull law gives a relationship between the probability F_K of fiber fracture and the applied stress (C). The evolution of $\ln[-\ln(1 - F_k)]$ according to $\ln C$ is called Weibull diagram.

The slope p of the curve $\ln C$ *versus* $\ln$ (where is the testing rate in µm/s) is related to the dynamic stress corrosion parameter n_d, a parameter characterizing the material capacity to resist to a stress. The accepted stress corrosion parameter is ~20 for high strength fiber.

Weibull plots corresponding to the fiber denoted by F1 (two polymer layer coating to protect the silica core) are given in Figure 3a.

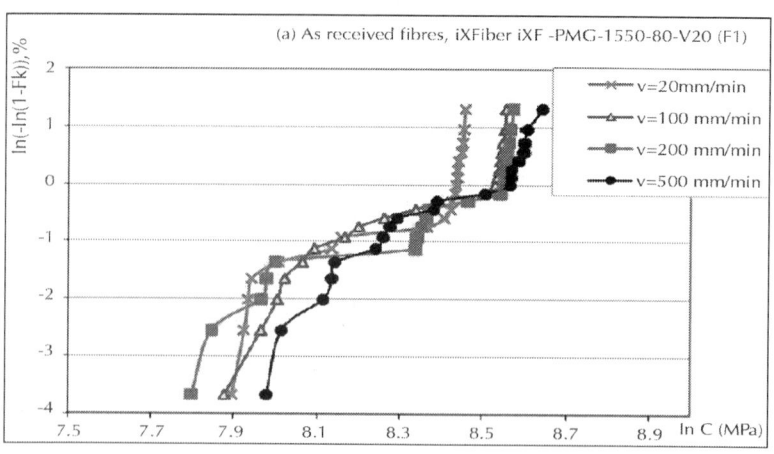

(a)

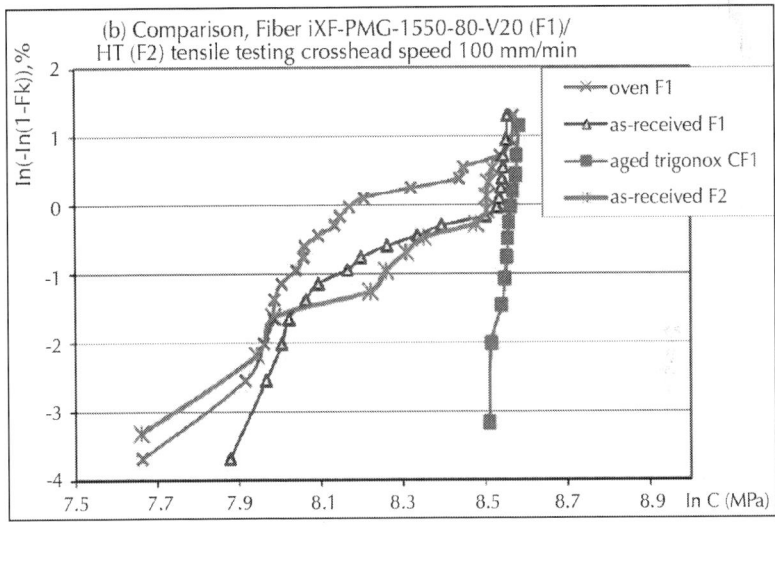

(b)

Figure 3: (a) Weibull plots for reference fiber F1 (double polymer coating) and (b) comparison of different etching factors tensile strength (F2—single polymer coating).

The medium values for the tensile stress correlated to the given strain rates were respectively 4034.3, 4183.7, 4328.5 and 4426.6 MPa. As compared to our previous dynamic tensile testing performed on Verrillon Inc. (North Grafton, MA, USA) and Alcatel optical fibers, one may conclude that F1 fiber presents quite broad dispersion and lower tensile stress. The stress corrosion factor n_d = 33.8 appears adequate with a linear interpolation of R^2 = 0.98.

Due to experimental reasons, the F2 fiber (single polymer layer coating) was tested for one speed (100 mm/min) and quite similar results are obtained, as seen in Figure 3b, with a medium stress of 4179.4 MPa, as compared to 4183.7 MPa for F1. The same broad dispersion and low tensile stress are reported.

The fabrication parameters influence the optical fiber tensile strength as seen in Figure 3b, thus heating in oven for 90 min at 80 °C led, as expected, to strength decrease, but the Trigonox C (resin compound) determined the strength increase with a more coherent and steeper distribution. Several testing performed with fiber F1 plunged for 7 days in water at room temperature (20 °C) confirmed our previous observations concerning slight increase of strength, but limited dispersion.

Pull-Out Testing—Interface Adhesion Characterization

The fabricated systems OF (F1)/Resin and OF (F1, F2)/Composite were prepared as described. The notch at 8 mm length from the sample end was cut to control the optical fiber embedding length L_e. The force (in N) *versus* displacement (in mm) curve was recorded and the maximum value was considered for comparison.

The debonding force F_d was taken as the maximum force preceding partial debonding.

In the case of optical fibers embedded in resin, noted OF (F1)/Resin, a poor interface adhesion is reported for all samples with a maximum of 0.98 N. Using the linear model Equation (1), an interfacial adhesion stress of 0.46 MPa is reported. The pull-out curves are similar to Figure 4.

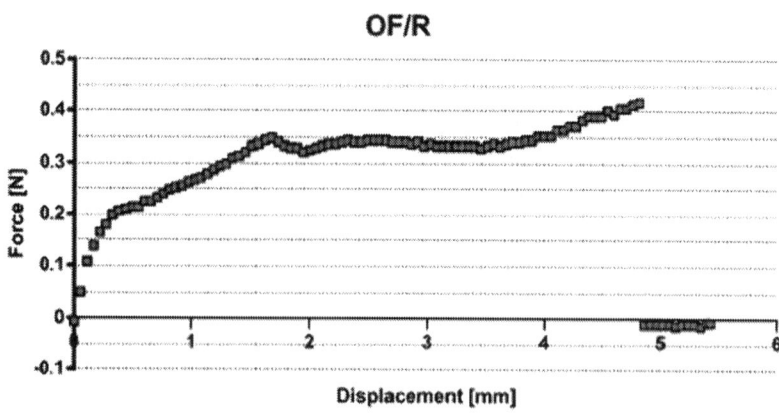

Figure 4: Pull-out record for optical fiber F1 embedded in epoxy-vinyl resin.

The optical fiber F1 embedded in the composite presented poor adhesion interface, too, slightly higher than the resin, but still low. A maximum value of 1.32 N (see Figure 5a) was recorded that corresponds to an interfacial adhesion stress of 0.66 MPa. but in the most cases debonding forces of 0.8–0.9 N were registered (see Figure 5b), leading to similar stresses as for non-reinforced matrix case.

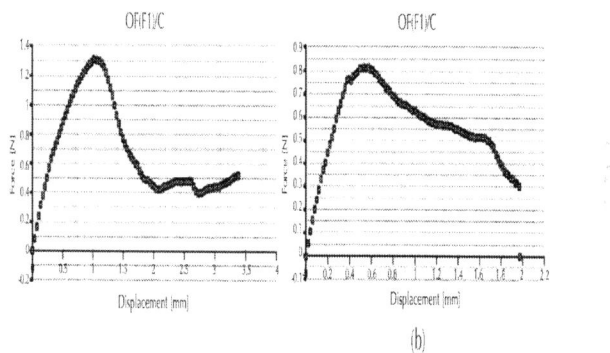

Figure 5: Pull-out record for optical fiber F1 embedded in epoxy composite/ glass fiber. (a) Maximum 1.32 N and (b) maximum 0.8 N.

Finally, the optical fiber F2 embedded in the composite presented better adhesion at OF/composite interface (Figure 6).

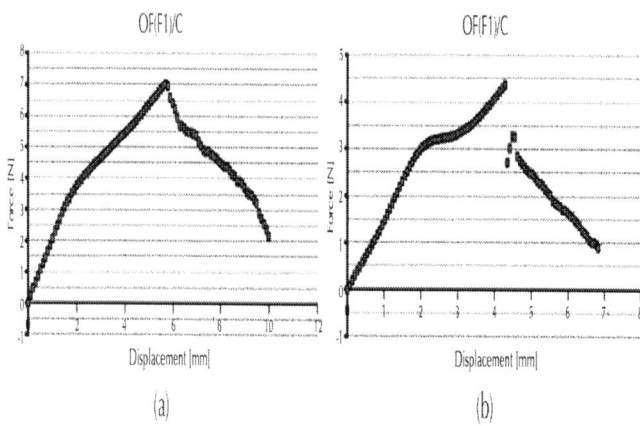

Figure 6: Pull-out record for optical fiber F2 embedded in epoxy composite/ glass fiber. (a) Maximum 7.1 N and (b) maximum 3.5 N.

A maximum value of 7.1 N (see Figure 6a) was recorded that corresponds to an interfacial adhesion stress of 3.53 MPa and several stick-slip phenomena were noticed (see Figure 6b), as previously reported [15], but for lower debonding forces.

The decrease of the force seems to be controlled by friction over the entire embedded length. The composite rigidity is higher than that of the non-reinforced resin. On the other hand, for the composite sample reinforced with glass fibers, one may expect less friction at the interface between the optical fiber and the composite than for non-reinforced resin in contact with the optical fiber coating. Thus, in the case of the composite sample with glass fibers, the interfacial debonding can propagate with a small instability and the obtained curve appears quasi-continual [15].

SEM Observations—Interface Adhesion Characterization

An overall image of the composite end is given in Figure 7 with the detail of the optical fiber. The single polymer coating may be seen in this case of F2 optical fiber centrally embedded in the composite, as compared to the double polymer coating in the case of F1 optical fiber (Figure 8a).

(a) (b)

Figure 7: (a) Micrograph of the optical fiber F2 embedded in the glass re-inforced vinyl-ester composite; (b) detail of the optical fiber (see core and single polymer layer).

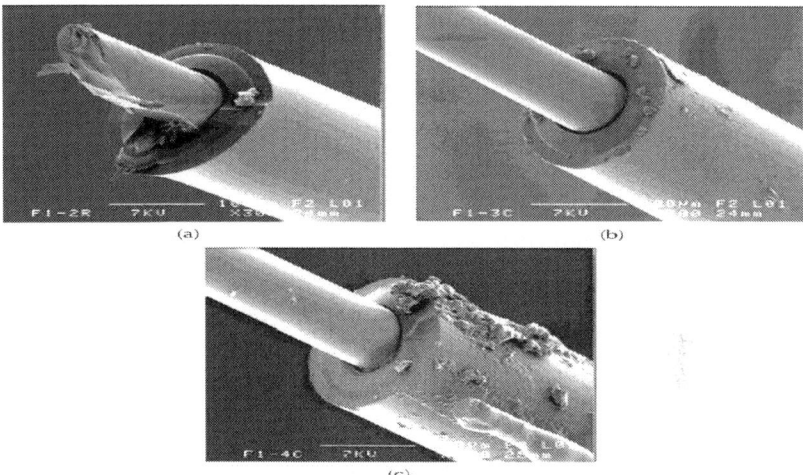

Figure 8: Interface fracture in F1 fiber embedded in (a) resin and in (b,c) composite.

In the case of the optical fiber (F1—double polymer coating) embedded in the resin, Figure 8a, examined after the pull-out test, the external polymer coating presented few resin matrix traces, but following the polymerization process the silica core interface appeared detached.

The physico-chemical nature of interfacial bonds depends on both elaboration temperature, and the time maintained at this elaboration temperature. The kinetics of interfacial bonding creation can be connected to a diffusive mechanism. The curing time enhances the formation of chemical bonds, and improves the diffusion mechanism at the interface optical fiber/resin. The adhesion at the interface results from the interdiffusion of macromolecular chains between the two polymer surfaces (resin and coating), thus forming an interphase. This interdiffusion could originate from a good compatibility between the acrylate coating and the vinyl ester resin, (both have hydroxyl groups, able to ensure a good wettability) or it could be the result of a macromolecular mobility, which leads to a tangle by reptation [15,16].

At the silica/polymer coating interface the coherence appears affected, as seen in Figure 8b, and this polymer detachment around the silica seems to be responsible for the low interfacial adhesion stress in case of resin and composite embedding F1 fiber. As noted, the resin filaments on the optical fiber external coating appeared more frequent in the composite case, Figure 8c.

Despite attentive precautions and rigorous fabrication among composite prepared samples, one may conclude that, following the pull-out test, half of the samples presented as-stripped optical fiber core (Figure 9a) or polymer coating with silica core. Losses because of detached coatings were due to core fracture and damaged adherence of inner soft polymer coating (Figure 9b).

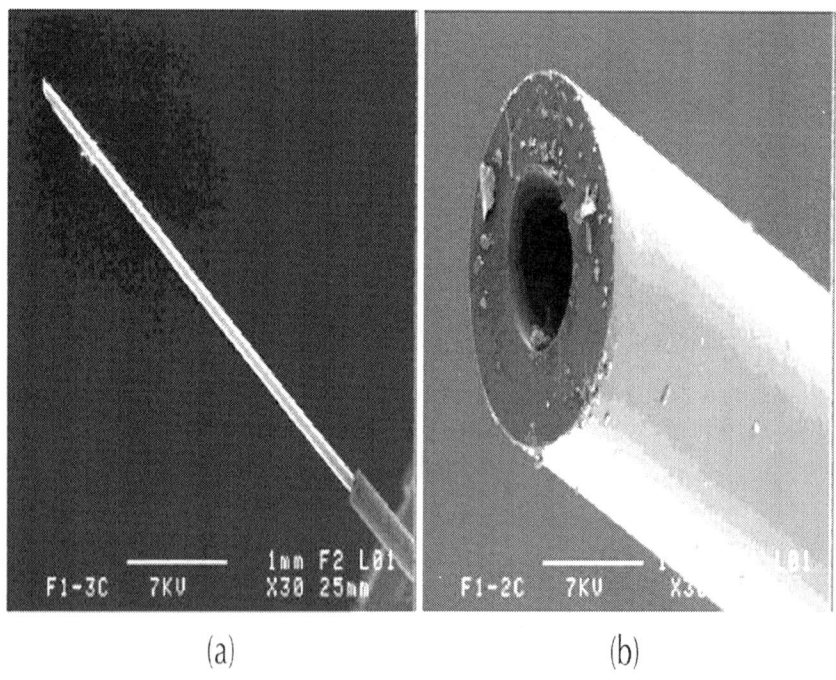

(a)   (b)

Figure 9: Micrograph of F1 fiber embedded in composite. (a) Core stripping; and (b) core damage.

For the F2 fiber embedded in the composite that has led to improved adhesion of the optical fiber, one may notice good cohesion at the silica core/polymer coating interface, see Figure 10a.

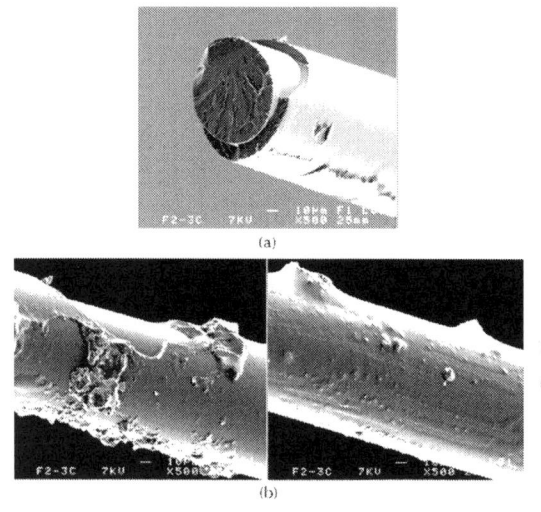

(a,b)

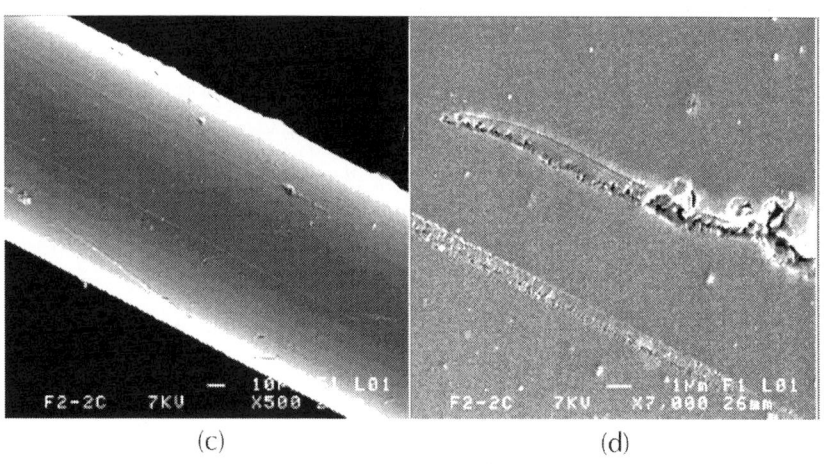

(c,d)

Figure 10: Micrograph of F2 fiber embedded in composite presents: (a) resin filaments; (b) polymer fracture, polymer detachment and glass reinforcements imprints; (c) scratches and (d) zoom of scratches.

In certain zones, maybe due to stress concentration, polymer fracture and detachment along the optical fiber (Figure 10b) was noticed, as were frequent resin traces (filaments and drops), sometimes oriented in the reinforcement direction.

Glass fiber reinforcements may scratch the polymer coating and certain imprints are also visible (Figure 10c), but this observation was not as frequent as previously reported [17].

SEM observations have confirmed that polymer detachment were more rare in the F2 fiber case than in the case of F1 fiber embedded similarly in the composite system.

CONCLUSIONS

As-received optical fibers F1 (double coating 172 μm diameter) and F2 (single coating 101.8 μm diameter) present similar mechanical strength with medium tensile strength between 4 and 4.4 GPa with broad dispersion and adequate stress corrosion factor ($n_d = 33.8$). Based on previous experimental studies, polymeric coatings are efficient to inhibit surface defects at the silica core interface.

The main difference between the investigated fibers is the thermal treatment applied in the case of F2 (single coating) that leads to a behavior close to polyimide.

As SEM observations exhibited, in the case of optical fiber with double layer polymer coating (F1), embedded in resin or composite, the soft polymer layer around the silica core retracted, probably due to the internal stresses following polymerization cycle of the resin matrix. Therefore, the interface has lost its coherence, leading to low interfacial adhesion stresses ranging from 0.4 to 0.66 MPa.

Optical fibers heat treatment at curing parameters revealed a strength decrease, but the resin catalytic system (Trigonox C) appeared favorable; the strength slightly increased and presented steeper distribution. Water at environment temperature has a similar effect after one week of exposure of the optical fiber, observations that appeared coherent to our previous results on other silica optical fibers

One may conclude that fabrication parameters acted synergistically on the optical fiber interfacial adhesion in the resin matrix, but certain precautions and a rigorous procedure should be followed to manage a coherent interface. Glass reinforcements of 55%–65%, fraction in mass, acted in the favor of the interfacial strength, the composite case exhibiting slightly better interfacial adhesion stress than the non-reinforced matrix resin, which may be explained through the harmful effect of internal stresses concentration in composite systems.

Finally, the optical fiber with a single layer coating (F2) subjected to a thermal treatment embedded in the composite system has allowed better quality at interface level. The polymer coated fiber appeared to be detached in certain zones following the pull-out test, scratched by reinforcements on its external surface and resin filaments. Drops were seen too, but the silica core/polymer coating revealed coherence and the interfacial adhesion stress is reported at 3.5 MPa.

Our objective to manage a reliable fabrication process of embedding optical fiber in an epoxy/glass fiber composite has been attainted and subsequent testing of Bragg gratings optical fiber (using the F2 single coating) should follow.

The authors express their gratitude to DFC-DCP (Creil, France) and to iXFiber S.A.S (Lannion, France) for technical assistance and for material support. A special remark for Joseph le Lannic from C.M.E.B.A. for MEB observations and fruitful discussions.

REFERENCES

1.   Davidson, R.; Roberts, S. Optical Fiber Sensor Compatibility and Integration in Composite Materials. In *Comprehensive Composite Materials*; Elsevier: Amsterdam, the Netherland, 2000; Volume 5, pp. 591–606.

2.   Berghmans, F.B. Reliability of optical fibers and components. *SPIE Proc* 2005, *5855*, 20–23.

3.   Wang, C.H.; Soufiane, A.; Majid, I.; Wei, K.; Drenzek, G. High reliability hermetic optical fiber for oil and gas application. *SPIE Proc* 2005, *5855*.

4.   Miao, P.; Kukureka, S.N.; Metje, N.; Chapman, D.N.; Rogers, C.D.F. Mechanical reliability of optical fibre for strain sensors. *SPIE Proc* 2005, *5855*, 1044–1047.

5.   Takeda, N. Structural health monitoring for smart composite structures systems in Japan. *Ann. Chim. Sci. Mater* 2000, *25*, 545–549.

6.   Takeda, N. Characterization of microscopic damage in composite laminates and real-time monitoring by embedded optical fiber sensors. *Int. J. Fatigue* 2002, *24*, 281–289.

7.   Zhou, G.; Sim, L.M. Evaluating damage in smart composite laminates using embedded EFPI strain sensors. *Opt. Lasers Eng* 2009, *47*, 1063–1068.

8.   Bettini, P.; di Landro, L.; Airoldi, A.; Baldi, A.; Sala, G. Characterization of the interface between composites and embedded fiber optic sensors or NiTiNOL wires. *Procedia Eng* 2011, *10*, 3490–3496.

9.   Gray, R.J. Analysis of the effect of embedded fiber length on fiber debonding and pull-out from an elastic matrix. *J. Mater. Sci* 1984, *19*, 861–870.

10.   DiFrancia, C.; Ward, T.C.; Claus, R.O. The single-fibre pull-out test, 1: Review and interpretation.*Compos. Part Appl. Sci. Manuf* 1996, *27*, 597–612.

11.   Poulain, M.; El Abdi, R.; Severin, I. Aging and Reliability of Single Mode Silica Optical Fibers. In*Optical Fibers Research Advances*; Schlesinger, J.C., Ed.; Nova Publishers: Hauppauge, NY, USA, 2007; pp. 355–368.

12.   Kukureka, S.N.; Cairns, D.R. Statistical analysis for strength distributions in optical fibres. *SPIE Proc*1999, *3848*, 17–27.

13.   Kukureka, S.N.; Cairns, D.R. Comparison of the distribution of estimated weibull parameters from optical fiber strength measurements and monte carlo simulations. *SPIE Proc* 2001, *4215*, 90–97.

14.   Mirer, T.; Ingman, D.; Zeifman, M. Non-equilibrum physical systems approach to modeling evolution of optical fiber reliability. *Phys. Lett* 2007, *365*, 181–186.

15.   Felcher, G.P.; Karim, A.; Russell, T.P. Interdiffusion at the interface of polymeric bilayers: Evidence for reptation. *J. Non-Cryst. Solids* 1991, *131–133*, 703–708.

16.   Bucknall, D.G.; Higgins, J.S.; Butler, S.A. Early stages of oligomer–polymer diffusion. *Chem. Eng. Sci* 2001, *56*, 5473–5483.

17.   Chean, V.; Matadi Boumbimba, R.; El Abdi, R.; Sangleboeuf, J.C.; Casari, P.; Drissi Habti, M. Experimental characterization of

interfacial adhesion of an optical fiber embedded in a composite material. *Int. J. Adhes. Adhes* 2013, *41*, 144–151.

Analysis of Seismic Damage of Underground Powerhouse Structure of Hydropower Plants Based on Dynamic Contact Force Method

Yang Yang [1, 2] Juntao Chen [1, 2], and Ming Xiao[1, 2]

[1]State Key Laboratory of Water Resources and Hydropower Engineering Science, Wuhan University, Wuhan 430072, China
[2]Key Laboratory of Rock Mechanics in Hydraulic Structural Engineering, Wuhan University, Ministry of Education, Wuhan 430072, China

ABSTRACT

Based on the characteristics of the dynamic interaction between an underground powerhouse concrete structure and its surrounding rock in a hydropower plant, an algorithm of dynamic contact force was proposed. This algorithm enables the simulation of three states of contact surface under dynamic loads, namely, cohesive contact, sliding contact, and separation. It is suitable for the numerical analysis of the dynamic response of the large and complex contact system consisting of underground powerhouse concrete structure and the surrounding rock. This algorithm and a 3D plastic-damage model were implemented in a dynamic computing platform, SUCED, to analyze the dynamic characteristics of the underground powerhouse structure of Yingxiuwan Hydropower Plant. By comparing the numerical results and postearthquake investigations, it was concluded that the amplitude and duration of seismic waves were the external factors causing seismic damage of the underground powerhouse structure, and the spatial variations in structural properties were the internal factors. The existence of rock mass surrounding the underground powerhouse was vital to the seismic stability of the structure. This work provides the theoretical basis for the anti-seismic design of underground powerhouse structures.

INTRODUCTION

The southwest region is the key area for hydropower development in China during the past few decades. A number of large-scale underground powerhouses of hydropower plants have been built in this region. It is also an earthquake-prone zone, with the seismic intensity usually above VII. Therefore, the underground powerhouses should have sufficient earthquake resistance capability which is vital to ensure normal operation of powerhouse and the safety of powerhouse personnel. Postearthquake investigations of the underground powerhouses in the epicentral region of Wenchuan earthquake, such as in Yingxiuwan, Yuzixi, and Gengda hydropower plants, reveal that underground powerhouses have, in general, stronger earthquake-resistant capability compared to their surface counterparts. Surrounding rock of the three underground powerhouses was generally stable, while the powerhouse concrete structure experienced obvious local cracking and failure. Concrete structure behaved as a weak link to affect the earthquake resistance capability of the underground powerhouses. Thus, it is very important to study the characteristics of seismic damage of underground powerhouse concrete structure.

It has been recognized that numerical method is an increasingly popular and effective strategy to solve the problem of structural dynamic response. A variety of works [1–5] have been devoted to the study of the dynamic response of underground powerhouse using the numerical method. Even though great achievement has been made in the field, the research on dynamic characteristic of powerhouse concrete structure is still lacking. From the microscopic point of view, the seismic damage and failure of underground powerhouse concrete structure are governed microscopic crack generation and propagation. In this way, the continuum method [6–8] (such as the finite element method, the finite difference method, and the boundary element method) and the discrete particle simulation method [9–22] (such as the discrete element method, the lattice-solid method, and the contact dynamics method) are the natural choice to simulate the damage and failure process. Among them, the finite element method is by far the most common and useful numerical method. The integration scheme it employs makes it appropriate for the widest variety of geologic and structural problems, and it can handle the most sophisticated constitutive relationships [23–25].

Underground powerhouse could be regarded as a system composed of powerhouse concrete structure and surrounding rock. Surrounding rock provides the external support for powerhouse concrete structure. As the concrete structure comes into direct contact with surrounding rock, seismic wave is propagated through rock mass to the powerhouse concrete structure. Therefore, the complex consisting of powerhouse concrete structure and surrounding rock undergoes a forced vibration. The simulation and analysis of the dynamic contact surface between surrounding rock and concrete structure is the key to the dynamic calculation of underground powerhouse concrete structure using the finite element method.At present, the numerical calculation methods for dynamic contact problems are primarily the Lagrange multiplier method [26], penalty method [27, 28], and their improved versions [29, 30]. However, these methods tend to either increase the degree of freedom of the system or influence the time step of integration [31, 32]. These methods adversely affect the precision and speed of calculation when applied to the analysis of an underground powerhouse concrete structure, which involves a large number of contact elements and complex contact states. Liu et al. [33, 34] put forward the dynamic contact force method targeted at the dynamic response problem of the contact crack. The convergence and stability of this algorithm were easy to meet, making it suitable for large and complex contact systems. But it cannot reflect the bond-slip properties of the contact surface.

In this paper, a new dynamic contact force method, considering the bond-slip properties of the contact surface, is suggested using the fundamental integration formulation of the dynamic contact force method. The algorithm of the method considers the cohesive effect of the contact surface between concrete structure and the surrounding rock. It is capable of simulating the large slip phenomenon of the contact surface under dynamic loads. Based on the proposed algorithm, a finite element model considering dynamic constitutive properties of materials is built for the dynamic numerical analysis of the underground powerhouse structure of Yingxiuwan Hydropower Plant. Then the characteristics of seismic damage of the underground powerhouse structure under dynamic loads are studied by means of the numerical analysis and postearthquake investigations. The results will provide a theoretical basis for the antiseismic design of underground powerhouse.

DYNAMIC CONTACT FORCE METHOD CONSIDERING THE BOND-SLIP PROPERTIES OF CONTACT SURFACE

The contact model of underground powerhouse concrete structure and surrounding rock is shown in Figure 1. Before application of the dynamic loading, the nodes on the contact surface between concrete and surrounding rock belong to point-to-point contact. A certain amount of cohesive force exists between the nodes, which are in cohesive contact state. During dynamic loading process, the stress in some contact nodes would exceed the cohesive force and enter into the state of sliding contact or even separation. When a large relative sliding occurs between these contact nodes, they would come into contact with surfaces of the adjacent elements and then belong to point-to-surface contact.

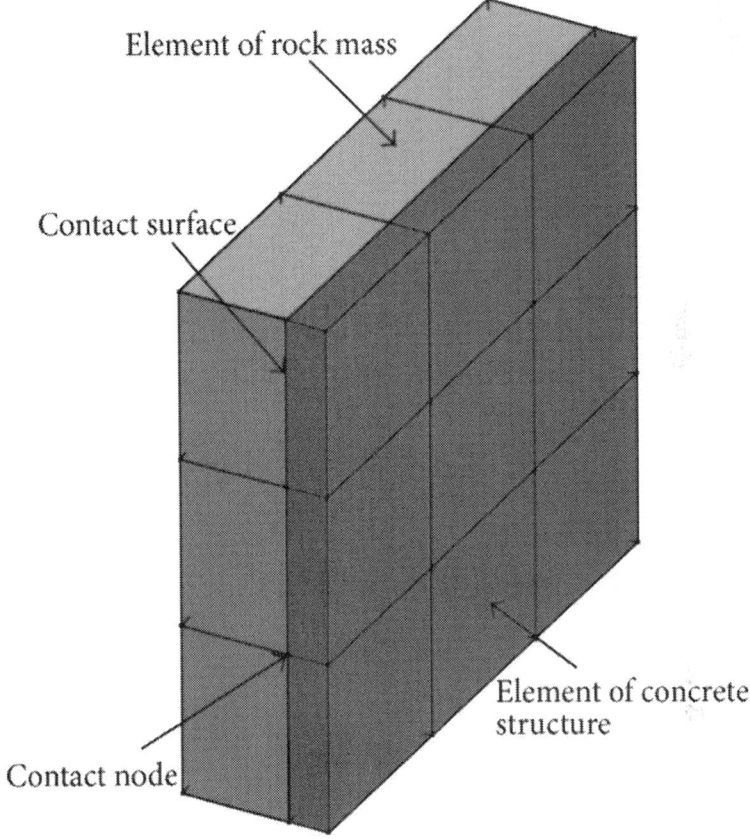

Figure 1: Contact model of concrete structure and surrounding rock.

Fundamental Integration Formulation of the Dynamic Contact Force Method

According to the dynamic contact force method proposed by Liu et al. [33, 34], the problem of dynamic response of structure containing interface is discretized using the finite element method. Then, the differential equations of the system could be obtained as follows:

$$M\ddot{U} + C\dot{U} + KU = F + R,$$

(1)

where M, C, and K are the mass, damping, and stiffness matrix, respectively. U is the displacement vector U is the velocity vector. U is the acceleration vector. F is the known vector of external forces. R is the dynamic contact force vector.

The central difference method is used to solve the differential equation at a time t. The time domain integral equation of the displacement and velocity of contact nodes containing the term of dynamic contact force could be obtained as follows:

$$\mathbf{U}^{t+\Delta t} = \overline{\mathbf{U}}^{t+\Delta t} + \frac{\Delta t^2}{2}\mathbf{M}^{-1}\mathbf{R}^t, \tag{2}$$

$$\dot{\mathbf{U}}^{t+\Delta t} = \frac{2\left(\mathbf{U}^{t+\Delta t} - \mathbf{U}^t\right)}{\Delta t} - \dot{\mathbf{U}}^t, \tag{3}$$

$$\overline{\mathbf{U}}^{t+\Delta t} = \left(1 - \frac{\Delta t^2}{2}\mathbf{M}^{-1}\mathbf{K}\right)\mathbf{U}^t$$

$$+ \left(\Delta t - \frac{\Delta t^2}{2}\mathbf{M}^{-1}\mathbf{C}\right)\dot{\mathbf{U}}^t + \frac{\Delta t^2}{2}\mathbf{M}^{-1}\mathbf{F}^t, \tag{4}$$

where t is the time step

R^t depends on the state of motion not only at time t but also at time $t+\Delta t$. Therefore, $U^{t+\Delta t}$ and $\dot{U}^{t+\Delta t}$ cannot be obtained directly by using (2)~(4). In order to obtain the motion state of the contact node at $t+\Delta t$, the contact force should be solved according to the contact conditions.

The Solution of Dynamic Contact Force under Point-to-Point Contact Condition

If relative sliding does not occur between contact nodes, or the relative sliding is small, the contact nodes are in or approximately in point-to-point contact condition. As shown in Figure 2, the node (i) on the contact surface of concrete and the node (i′) on the contact surface of the surrounding rock are in point-to-point contact at time t.

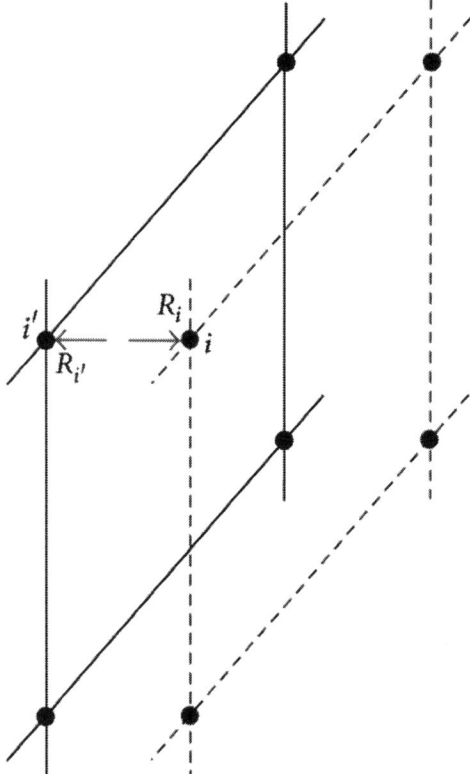

Figure 2: Relationship of point-to-point contact.

Suppose the contact node pair at time $t+ t$ is in cohesive contact state; then in the normal and tangential direction, the contact nodes pair should meet the nonintrusive condition and the displacement compatibility condition with no relative sliding, respectively, as follows:

$$\left[\left(\mathbf{U}_{i'}^{t+\Delta t} - \mathbf{U}_{i}^{t+\Delta t}\right)\mathbf{n}_{i}\right]\mathbf{n}_{i} = 0,$$

$$\left(\mathbf{U}_{i}^{t+\Delta t} - \mathbf{U}_{i'}^{t+\Delta t}\right) - \left[\left(\mathbf{U}_{i}^{t+\Delta t} - \mathbf{U}_{i'}^{t+\Delta t}\right)\mathbf{n}_{i}\right]\mathbf{n}_{i}$$

$$= \left(\mathbf{U}_{i}^{t} - \mathbf{U}_{i'}^{t}\right) - \left[\left(\mathbf{U}_{i}^{t} - \mathbf{U}_{i'}^{t}\right)\mathbf{n}_{i}\right]\mathbf{n}_{i},$$

$$(5)$$

where $\mathbf{n}_{i}$ is the unit normal vector of contact node.

Equation (2) is substituted into (5). According to the principle that a pair of dynamic contact force is equal in magnitude but opposite in direction, that is, R_i^t $R_{i'}^t$, we have

$$N_i^t = \frac{2M_i M_{i'}}{(M_i + M_{i'})\,\Delta t^2}\,\Delta_1,$$

(6)

$$T_i^t = \frac{2M_i M_{i'}}{(M_i + M_{i'})\,\Delta t^2}\,\Delta_2,$$

(7)

where N_i^t, T_i^t, are the normal and tangential components of , and $_1$, $_2$, and satisfy the following equations, respectively:

$$\Delta_1 = \left[\left(\overline{U}_{i'}^{t+\Delta t} - \overline{U}_i^{t+\Delta t}\right)n_i\right]n_i,$$

(8)

$$\Delta_2 = \overline{U}_{i'}^{t+\Delta t} - \overline{U}_i^{t+\Delta t} + U_i^t - U_{i'}^t$$
$$- \left[\left(\overline{U}_{i'}^{t+\Delta t} - \overline{U}_i^{t+\Delta t} + U_i^t - U_{i'}^t\right)n_i\right]n_i,$$

(9)

$$N_i^t = \left(R_i^t n_i\right)n_i,$$

(10)

$$T_i^t = R_i^t - \left(R_i^t n_i\right)n_i.$$

(11)

In the above equations, T_i^t and N_i^t were obtained from the analysis of motion of the node pair. Therefore, they must satisfy the following inequalities.

$$\left\|\mathbf{T}_i^t\right\| \leq \mu_s \left\|\mathbf{N}_i^t\right\| + cA, \quad \left\|\Delta_1\right\| \geq 0.$$

(12)

If the value of $\mathbf{T}_i^t$ does not satisfy (12), then the node pair would enter into the state of sliding contact. We have

$$\mathbf{T}_i^t = \mu_d \left\|\mathbf{N}_i^t\right\| \frac{\mathbf{T}_i^t}{\left\|\mathbf{T}_i^t\right\|}.$$

(13)

$$\sqrt{\left(\mathbf{T}_i^t\right)^2 + \left(\mathbf{N}_i^t\right)^2} \leq cA, \quad \left\|\Delta_1\right\| < 0$$

(14)

If the value of $\mathbf{T}_i^t$ does not satisfy (14), then the node pair entered into the separation state. We have

$$\mathbf{T}_i^t = 0, \qquad \mathbf{N}_i^t = 0,$$

(15)

where μ_s and μ_d are the coefficients of static friction and kinetic friction, respectively A is the control area of the contact node to be calculated. c is the cohesive force between the contact surface. Before time $t + t$, if sliding or separation has not occurred between the node pair, $c > 0$, otherwise, $c = 0$.

The Solution of Dynamic Contact Force under Point-to-Surface Contact Condition

If larger relative sliding has occurred between contact nodes under dynamic loads, the contact nodes will be in the state of point-to-surface contact. Then there are no cohesive forces between the contact nodes and surfaces. As shown in Figure 3, at time t, one node i on the contact surface of concrete structure (or the surrounding rock) comes into contact with the surface of the surrounding rock (or concrete structure). Such a contact point on the contact surface is denoted as. i'

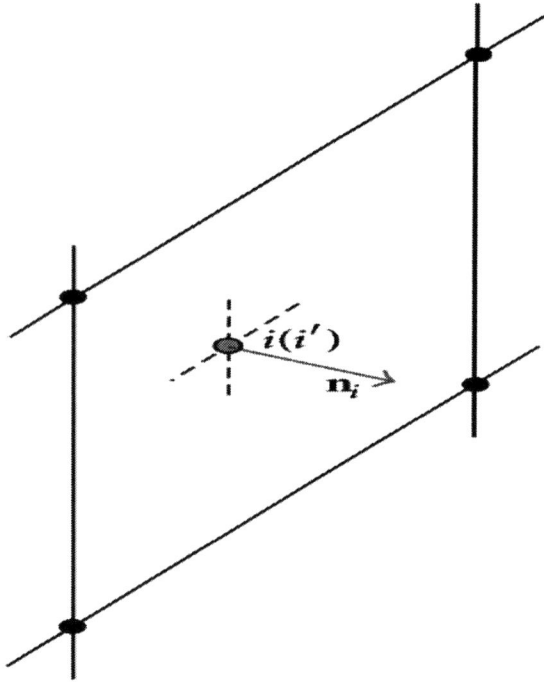

Figure 3: Relationship of point-to-surface contact.

It is assumed that node i is in the state of cohesive contact with the corresponding contact surface at time $t+ t$. In the normal and tangential direction, the node should meet the nonintrusion condition and the displacement compatibility condition with no relative sliding, respectively, represented by (5). The displacements of the contact point i' at time t and time $t+ t$ are, respectively, given by

$$\mathbf{U}_{i'}^{t} = \sum_{j}\phi_{j}\mathbf{U}_{j}^{t}, \qquad \mathbf{U}_{i'}^{t+\Delta t} = \sum_{j}\phi_{j}\mathbf{U}_{j}^{t+\Delta t},$$

(16)

where ϕ_{j} is the shape function. j is the node number of the contact surface.

Substituting (16) and (2) into (5), we have

$$\frac{\Delta t^2}{2M_i}\mathbf{N}_i^t - \sum_j \frac{\Delta t^2 \phi_j}{2M_j}\mathbf{N}_j^t = \Delta_3,$$

(17)

$$\frac{\Delta t^2}{2M_i}\mathbf{T}_i^t - \sum_j \frac{\Delta t^2 \phi_j}{2M_j}\mathbf{T}_j^t = \Delta_4,$$

(18)

Where

$$\Delta_3 = \left[\left(\sum_j \phi_j \overline{\mathbf{U}}_j^{t+\Delta t} - \overline{\mathbf{U}}_i^{t+\Delta t}\right)\mathbf{n}_i\right]\mathbf{n}_i,$$

$$\Delta_4 = \sum_j \phi_j \overline{\mathbf{U}}_j^{t+\Delta t} - \overline{\mathbf{U}}_i^{t+\Delta t} + \mathbf{U}_i^t - \sum_j \phi_j \mathbf{U}_j^t$$

$$-\left[\left(\sum_j \phi_j \overline{\mathbf{U}}_j^{t+\Delta t} - \overline{\mathbf{U}}_i^{t+\Delta t} + \mathbf{U}_i^t - \sum_j \phi_j \mathbf{U}_j^t\right)\mathbf{n}_i\right]\mathbf{n}_i.$$

(19)

In the above equations, two conditions should be discussed.

- If $\|\Delta_3\| < 0$, node i is separated from the corresponding contact surface. $\mathbf{N}_i^t$, $\mathbf{T}_i^t$ can be computed from (15).
- If $\|\Delta_3\| \geq 0$, node i is in contact with the corresponding contact surface. It is assumed that (17) will result in $\|\Delta_3\| \geq 0$ for m-number of nodes. Thus, (17) and (18) of the nodes can be represented as follows:

$$[\mathbf{H}]_{m\times m}\left[\mathbf{N}^t\right]_{1\times m} = [\Delta_3]_{1\times m},$$

(20)

$$[\mathbf{H}]_{m\times m}\left[\mathbf{T}^t\right]_{1\times m} = [\Delta_4]_{1\times m},$$

(21)

where $[H]_{m \times m}$ is the coefficient matrix, relating $t, M,$ and ϕ.

Solving (20) and (21), the contact forces N^t and T^t can beobtained for the nodes in point-to-surface contact condition.If one node has $\|T^t\| > \mu_s\|N^t\|$, then the node enters intothe state of sliding contact. For this particular node, T^tcan becomputed from (13), and (18) of the node should be removedfrom (21). Solving the new equation (21), the tangentialcontact force of other nodes can be known.

$$R^t = N^t + T^t. \tag{22}$$

Then, the displacement and velocity of each contact node for the next time step can be computed from (2) and (3). Figure 4 presents the flow chart for the proposed method.

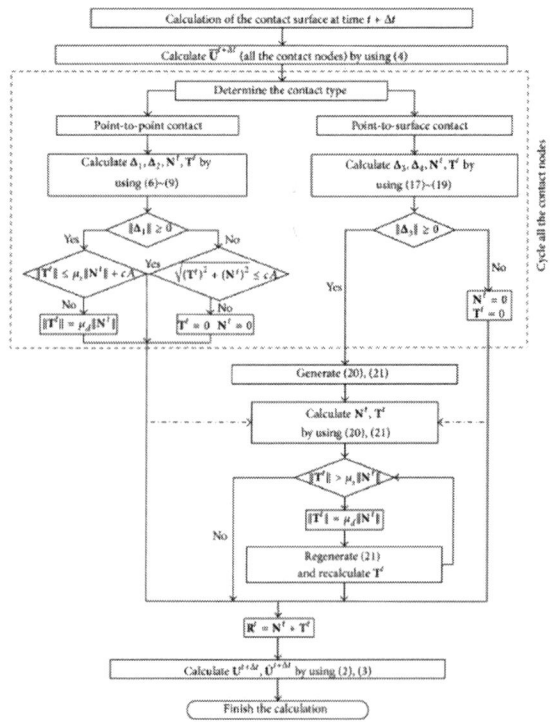

Figure 4: The flow chart for the dynamic contact force method.

DYNAMIC CONSTITUTIVE MODEL OF CONCRETE AND ROCK MASS

Under cyclic loading, the unloading stiffness of concrete and rock mass at the later yielding stage is lower than the stiffness at the initial linear stage. The plastic-damage model, proposed by Lubliner et al. [35], improved by Lee and Fenves [36, 37], and generalized for 3D model by Omidi and Lotfi [38], could effectively simulate such a phenomenon. The model is suitable for the dynamic analysis of the quasibrittle materials, such as concrete and rock mass [3].

According to the basic theory of plastic-damage model, the plastic-damage stress-strain relationship of rock mass or concrete can be expressed as follows:

$$\sigma = (1 - D)\,\overline{\sigma},$$

$$\overline{\sigma} = E_0 : \left(\varepsilon - \varepsilon^p\right),$$

(23)

where σ is the stress tensor. $\overline{\sigma}$ is the effective stress tensor. E_0 is the initial stiffness of the material. ε is the strain tensor. ε_p is the plastic strain tensor. D is the damage coefficient.

The structural damage is the result of the microcracks of the material. Under cyclic loading, the opening and closure of microcracks may happen, making the damage as a complex mechanism. When the state of stress especially changes from tensile to compressive, the stiffness weakened by the damage begins to recover. In order to simulate this phenomenon, the damage coefficient can be written as follows:

$$D = 1 - \left(1 - D_c\right)\left(1 - sD_t\right),$$

$$D_t = 1 - \exp\left(-d_t \varepsilon^p\right),$$

$$D_c = 1 - \exp\left(-d_c \varepsilon^p\right),$$

(24)

where D_t and D_c are tensile and compressive damage coefficients, respectively. d_t and d_c are the dimensionless constants as the functions of plastic strain. s $(0 \leq s \leq 1)$ is the coefficient of restitution when the material shifts from tensile state to compressive state.

The yield function of the model in form of effective stress is given as follow:

$$F\left(\bar{\sigma}, \varepsilon^P\right)$$

$$= \frac{1}{1-\alpha}\left[\alpha \bar{I}_1 + \sqrt{3\bar{J}_2} + \beta\left(\varepsilon^P\right)\left\langle \hat{\bar{\sigma}}_{\max}\right\rangle - \gamma\left\langle -\hat{\bar{\sigma}}_{\max}\right\rangle\right]$$

$$- \bar{\sigma}_c\left(\varepsilon^P\right),$$

(23)

where α and γ are the dimensionless constants and β is a constant variable. For more details one can consult Omidi and Lotfi [38]. I 1 and J 2 are the first and second invariants of the effective stress tensor. $\hat{\sigma}$ max is the maximum effective principal stress.

Concrete and the surrounding rock are used as friction material. The nonassociated flow rule can simulate the volume expansion properties under the compressive state. Therefore, the plastic-damage model employs nonassociated flow rule. The plastic potential function follows the Drucker-Prager hyperbolic function as follows:

$$\Phi\left(\bar{\sigma}\right) = \sqrt{\left(\xi\alpha_p f_{t0}\right)^2 + 2\bar{J}_2} + \alpha_p \bar{I}_1,$$

(26)

where ξ is the parameter which controls the potential function to approach the asymptote, α_p is the dilatancy parameter, and f t_0 is the maximum uniaxial tensile strength of the material.

An implicit cylindric anchor bar element method is adopted for the finite element implementation of the steel bars embedded in rock or concrete structures. The embedded steel bars are considered to improve the stiffness of the rock or concrete structures in this model. Therefore the stiffness of the steel bars can be superimposed onto the stiffness of rock or concrete element during numerical simulation. This method is detailed in [39].

SEISMIC ANALYSIS OF UNDERGROUND POWERHOUSE STRUCTURE OF YINGXIUWAN HYDROPOWER PLANT

The dynamic analysis of underground powerhouse structure was performed with an in-house FEM numerical simulation platform, SUCED [40], which was designed for estimating seismic damage of large underground caverns group. The platform is based on the dynamic time-history method and uses explicit central difference method to solve the finite element equation.

Postearthquake Investigations of Underground Powerhouse Structure of Yingxiuwan Hydropower Plant

Yingxiuwan Hydropower Plant is one of the nine cascade hydropower plants of Minjiang River. This plant has an underground powerhouse, with three generating units. The seismic intensity of Yingxiuwan Hydropower Plant is up to XI, according to the distribution of seismic intensity of Wenchuan earthquake. Located just 8 km away from the epicenter, Yingxiuwan is one of the hydropower plants closest to the epicenter. A survey performed after the earthquake revealed some important findings. (a) Existing cracks on the arch of the underground powerhouse structure had propagated. A number of cracks appeared along the river and along the axis of the powerhouse after the earthquake (as shown in Figure 5(a)). (b) Part of sidewall lining cracked, but the crack width and the length were limited (as shown in Figure 5(b)). (c) The floor of turbine and generator layer showed closed cracks. The overlying floor was uplifted (as shown in Figure 5(c)). (d) The concrete of the inside wall of generator pedestal experienced a large number of cracks (as shown in Figure 5(d)).

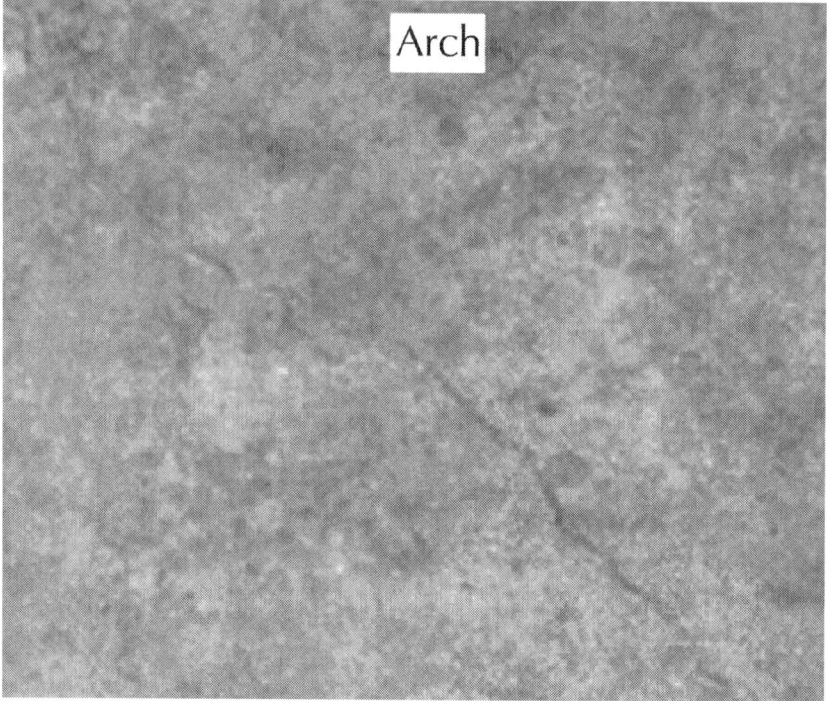

Figure 5: Destruction of powerhouse concrete structure.

Numerical Calculation Model

Three-dimensional finite element model including surrounding rock and concrete structure of the powerhouse was established. The model consisted of 189,187 nodes and 171,432 hexahedron elements (eight nodes). The powerhouse is 52.8 m long, 17.0 m wide, and 37.2 m high. The model ranges along x, y, and z-axes are 96.0, 88.2, and 147.7 m, respectively. In order to guarantee the accuracy of dynamic calculation results, the maximum mesh size should be less than 7.5 m based on the cut-off frequency of seismic waves. As the maximum mesh size of the model is 7.0 m, the demand of FEM dynamic simulation can be satisfied. The profile of the model is shown in Figure 6.

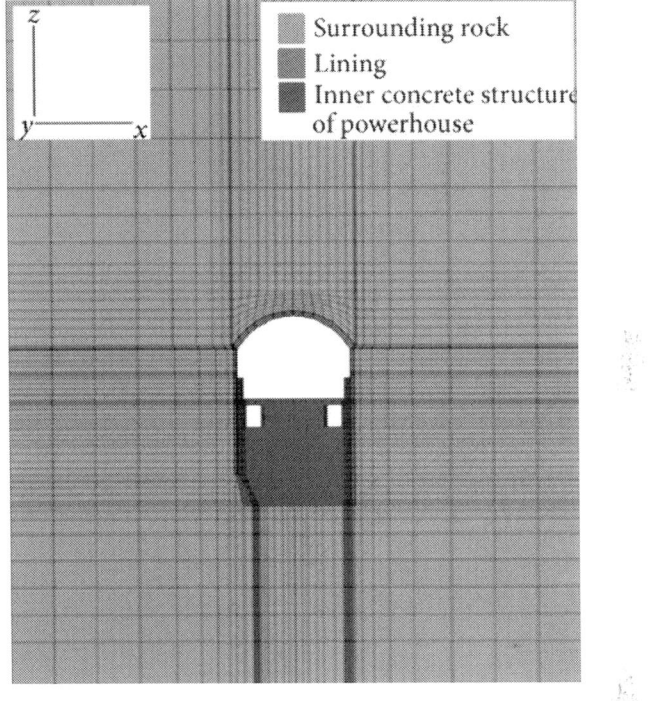

Figure 6: Profile of finite element model.

Mechanical Parameters of Material

The mechanical parameters of the rock mass and concrete material are shown in Table 1. The 3D plastic-damage model discussed in Section 3 was used for the numerical simulation of the surrounding rock and concrete structure.

Table 1: Mechanical parameter of rock mass and concrete

Material	Deformation modulus (GPa)	Poisson ratio	Cohesion (MPa)	Friction angle (o)	Tensile strength (MPa)
Rock	10	0.25	2.18	41.6	1.97

Concrete (lining and inner structure)	25	0.17	2.0	46	1.30

Boundary Conditions

Free field boundary condition was applied to absorb the reflection waves along the four vertical boundaries. Viscoelastic artificial boundary condition was used to absorb the incident waves from the bottom of the model [41].

Dynamic Loads

Wolong was the nearest strong motion station to the epicenter to record strong ground motion from the Wenchuan earthquake. Its epicentral distance was 19 km. The seismic waves recorded in Wolong were concluded more representative than those recorded in other stations. The 20 s to 80 s sections of Wolong monitored acceleration data, which have high intensity and amplitude, were used as the seismic input load. The input acceleration time-histories used in the calculation model are shown in Figure 7.

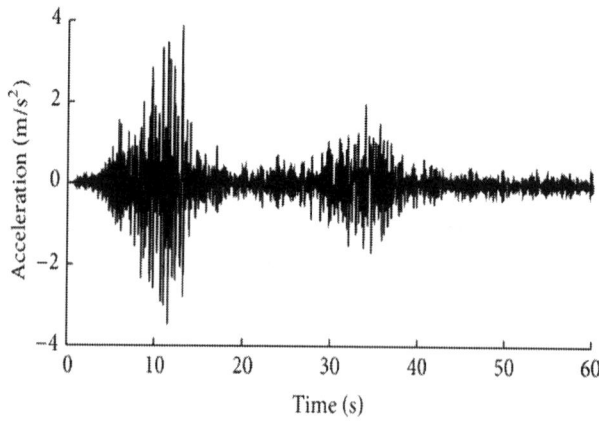

Figure 7: Input acceleration time-histories of calculation model.

Numerical Calculation Results

Static analysis of the surrounding rock excavation and concrete structure of the underground powerhouse was performed before performing a dynamic analysis. The results provided the initial conditions for a dynamic analysis.

A formula as introduced in (27) was used to calculate the damage volume of all damaged elements of underground powerhouse concrete structure:

$$V = \sum_{i=1}^{n} v_i^D,$$

(27)

where v is the total damaged volume of the concrete structure. V_i^D is the tensile or compressive damaged volume of each single element, and n is the number of damaged elements.

Analysis of Evolving Process of Seismic Damage

The time history of the total damaged volume of powerhouse structure was plotted as shown in Figure 8. As can be seen, the totaldamagedvolumewas63.2m³ before seismic loading. Seismic damage on the powerhouse structure initiated when $t = 2.6$ s. From $t = 2.6$ s to 20s and from$t = 30$ s to 40s, the seismic waves entered into the two peaks.The total damaged volume increased sharply. At other times, the amplitudes of the seismic waves were small. So the total damaged volume increased slowly. When $t = 60$ s, the total damaged volume was about 460.3m³.

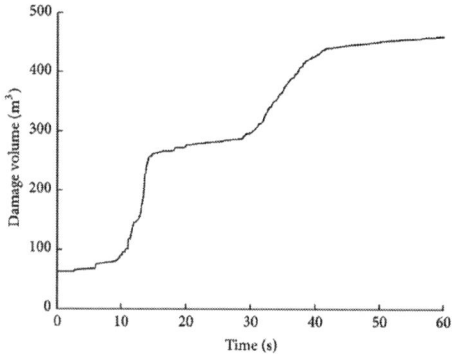

Figure 8: Time history of the variation of the seismic damaged volume of the powerhouse structure.

Figures 9 and 10 show the plot of the damage coefficient when $t = 0$s and $t = 60$ s, respectively. Before seismic loading, a small amount of damage zones distributed in the upper part of the lining. The damaged coefficient did not exceed 0.3. When the seismic loading was completed, a large amount of damage zones appeared in the lining. The damaged areas were mainly distributed in the arch and the upper sidewall. The damage coefficient at some locations was observed as high as 1.0. The concrete showed the risk of cracking failure.

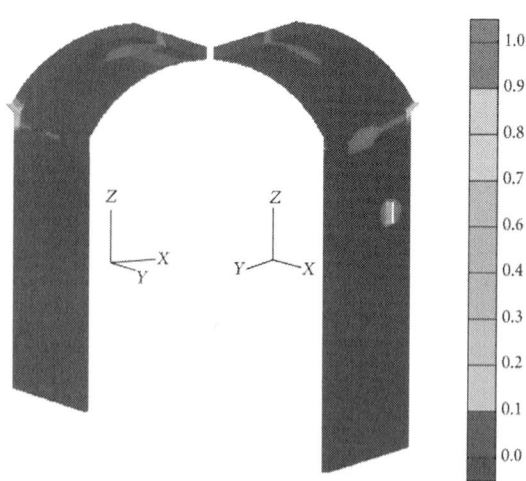

Figure 9: Damage coefficient distribution of the lining when $t = 0$ s.

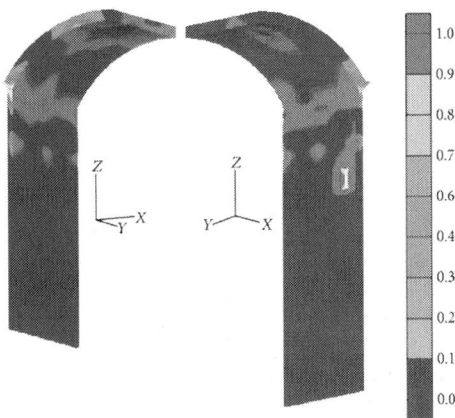

Figure 10: Damage coefficient distribution of the lining when $t = 60$ s.

The damage coefficient distribution of the inner concrete structure of the powerhouse when s and s was plotted (Figures 11 and 12). As can be seen, when the seismic loading was completed, a wide range of damaged areas appeared at the floor of turbine and generator layer and the inside wall of generator pedestal. The damage coefficient was close to 1.0 at most of the damaged area. The concrete showed the risk of cracking failure. The damaged areas were also noticed at the crane beam corbel and the column of turbine layer. However, the damage coefficient was not large.

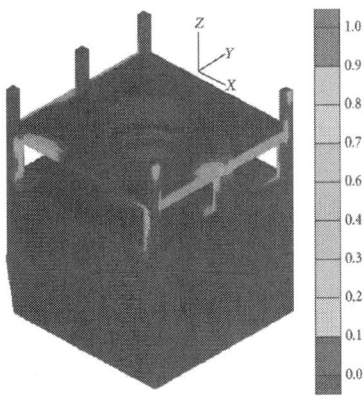

Figure 11: Damage coefficient distribution of the inner concrete structure when $t = 60$ s.

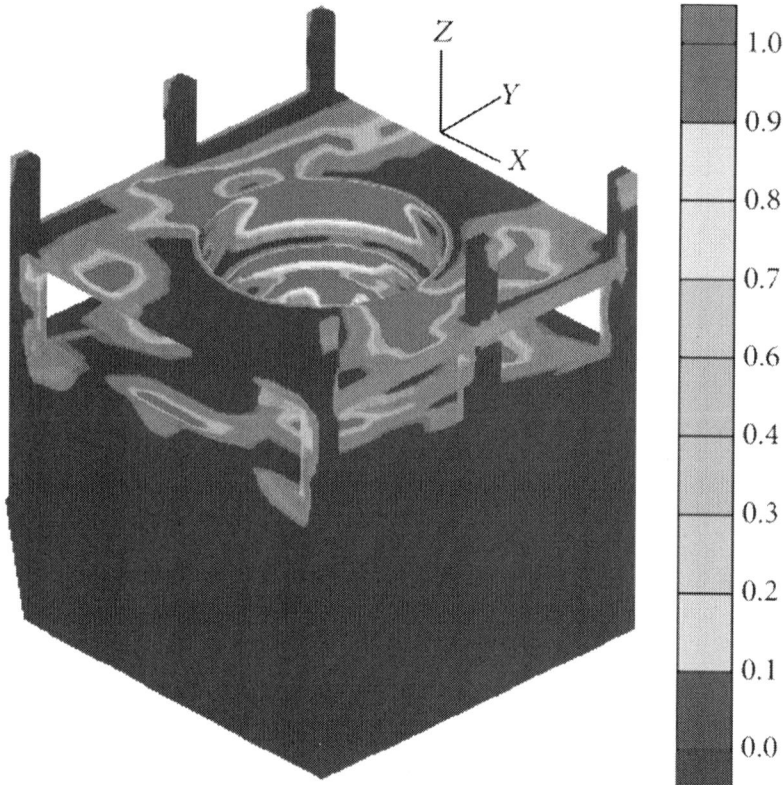

Figure 12: Damage coefficient distribution of the inner concrete structure when $t = 60$ s.

Analysis of the Sliding and Separation of Contact Surface

The distribution of the sliding and separation zone of the contact surface between concrete structure and surrounding rock when $t = 0$s and $t = 60$s was plotted as shown in Figures 13 and 14. When $t = 0$s, a small amount of sliding or separation zones occurred on the lining. When $t = 60$ s, the sliding or separation zones of the contact surface were greatly expanded, which mainly occurred on the arch, the junction of the sidewall with the arch, and the junction of caverns. This is consistent with the distribution of the damaged areas of the lining structure.

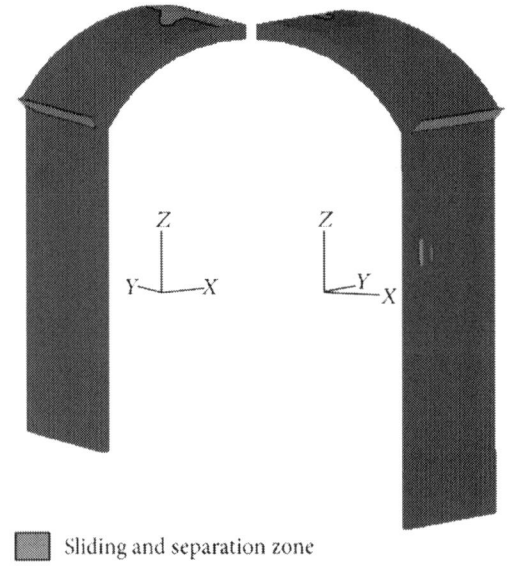

Sliding and separation zone

Figure 13: Distribution of sliding and separation zone when $t = 0$ s.

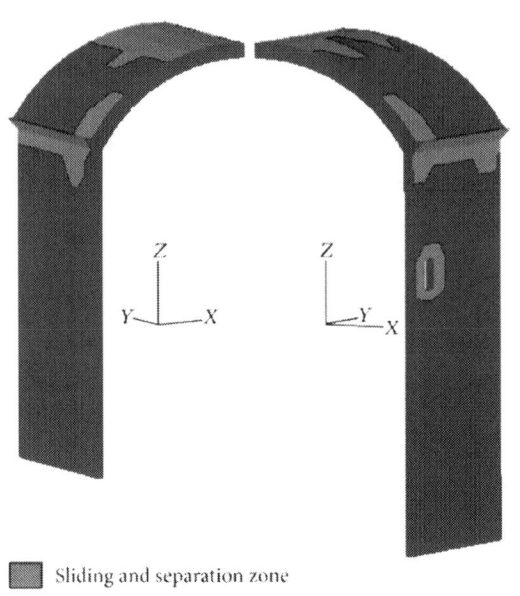

Sliding and separation zone

Figure 14: Distribution of sliding and separation zone when $t = 60$ s.

ANALYSIS OF SEISMIC DAMAGE CHARACTERISTICS OF UNDERGROUND POWERHOUSE STRUCTURE

Comparison between Numerical Calculation and Postearthquake Investigation

The numerical calculation results showed that the damaged areas of underground powerhouse concrete structure mainly occurred on the arch, the upper sidewall, the floor of generator and turbine layers, the inside wall of generator pedestal, and the junction of caverns. This is basically consistent with the cracking failure areas noticed in the postearthquake investigations. Therefore, the dynamic calculation method for underground powerhouse concrete structure proposed in this paper is proved to be reasonable and effective. The results could truly reflect the damage characteristics of powerhouse structure and could provide theoretical basis for antiseismic design of such structures.

Influence of Seismic Waves on Seismic Damage of the Structure

The degree of damage of the underground powerhouse structure was closely related to the amplitude and duration time of seismic waves. The larger the amplitude and the longer the duration of seismic waves were, the more obvious the damage of the underground powerhouse structure was. In the numerical simulation of the underground powerhouse structure of Yingxiuwan Hydropower Plant, the distribution range of the damaged areas and the damage coefficient increased significantly at the time of the two obvious vibration processes of seismic waves. When the two obvious vibration processes were over, the amplitude of the subsequent seismic waves was smaller. However, due to the continued input of seismic waves, the extent of damaged areas and the damage coefficient increased to a certain degree. This shows that the amplitude and duration time of seismic waves determine the degree of the damage of an underground powerhouse structure. These are indeed the external factors causing the seismic damage of the underground powerhouse structure.

Influence of Structural Properties on Seismic Damage of the Structure

The structure of the underground powerhouse possesses significant spatial variation. The structure above the generator layer is mainly the lining, which is subjected only to the unidirectional constraint of the surrounding rock. The upper structure is relatively weak. Under seismic loads, the damaged areas were distributed extensively. The structure below the generator layer includes beams, plates, columns, and other massive concrete structures, which are subject to the constraints of surrounding rock on all sides. The proportion of the damaged areas was relatively smaller compared with that of the upper structure. Nevertheless, a large number of damaged areas occurred at the beams, plates, columns, and the inner wall of the generator pedestal, which have lower structural strength. It is apparent that the spatial variation of underground powerhouse structure led to the varying degrees of damage. This is an internal factor causing the seismic damage of the underground powerhouse structure.

Influence of Surrounding Rock on Seismic Damage of the Structure

The postearthquake investigations in many other plants showed that the characteristic seismic damage of surface powerhouses was predominantly the horizontal shear failure. These damages were serious and difficult to repair. On the other hand, the damage type of underground powerhouse structure was mainly the surface shedding and closed cracks. The degree of damage was generally lighter. The results of the numerical analysis showed that the damage of the powerhouse structure in the areas which could slide relative to or separate from the surrounding rock was more serious while that in the areas having good contact with the surrounding rock was lighter. Hence, the constraint of the surrounding rock remained influential to maintain the stability of the underground powerhouse structure under seismic loads. This influence of the surrounding rock is the main reason for the lesser damage of an underground powerhouse structure compared to its surface counterpart.

Under seismic loads, the surrounding rock deformed inward. The concrete structure was subjected to compression caused by the deformation of the surrounding rock. It was more pronounced in the lower part of underground powerhouse structure which would be subjected to constraints on all sides. For the underground powerhouse structure of Yingxiuwan Hydropower Plant, the floor of the generator layer was uplifted under the action of compression imposed by the surrounding rock. Therefore, the surrounding rock may have both advantages and disadvantages for underground powerhouse structure. However, in general, the advantages outweigh the disadvantages.

DISCUSSION ON THE ANTISEISMIC MEASURES OF UNDERGROUND POWERHOUSE STRUCTURE

According to the seismic damage characteristics, the constraint imposed by the surrounding rock remained as an important factor to the antiseismic stability of the underground structure. The powerhouse structure above the generator layer suffered from the problem of "underconstraint," resulting in a large area of damage in the upper structure under seismic loads. For the structures below the generator layer, the problem of "overconstraint" made the action of compression caused by the surrounding rock deformation more obvious.

To ensure the antiseismic design, surrounding rock is recommended to be reinforced to improve its stability. In this way, the favorable constraint of the surrounding rock on the internal structure can be increased, while the unfavorable impact is minimized. Secondly, anchorage support or other support measures are recommended to strengthen the constraint of surrounding rock for the structure above the generator layer, so as to promote the antiseismic capacity of the upper structure. Finally, soft-seismic isolation layer can be added between the lower parts of the concrete structure and the surrounding rock. The soft-seismic isolation layer can not only weaken the compression caused by the surrounding rock deformation, but also reduce the seismic response of the structure to a certain extent. The enhancement in the antiseismic capability of the structure after incorporating the abovementioned measures was evident from the numerical simulation. For instance, Figures 15 and 16 show the distribution of the damaged areas in the powerhouse structure after adopting the abovementioned antiseismic measures. Significant improvement is seen in the structural performance when compared to the damage in Figures 10 and 12. The distribution range and the damage coefficient were greatly reduced. The stability of the powerhouse concrete structure was enhanced considerably.

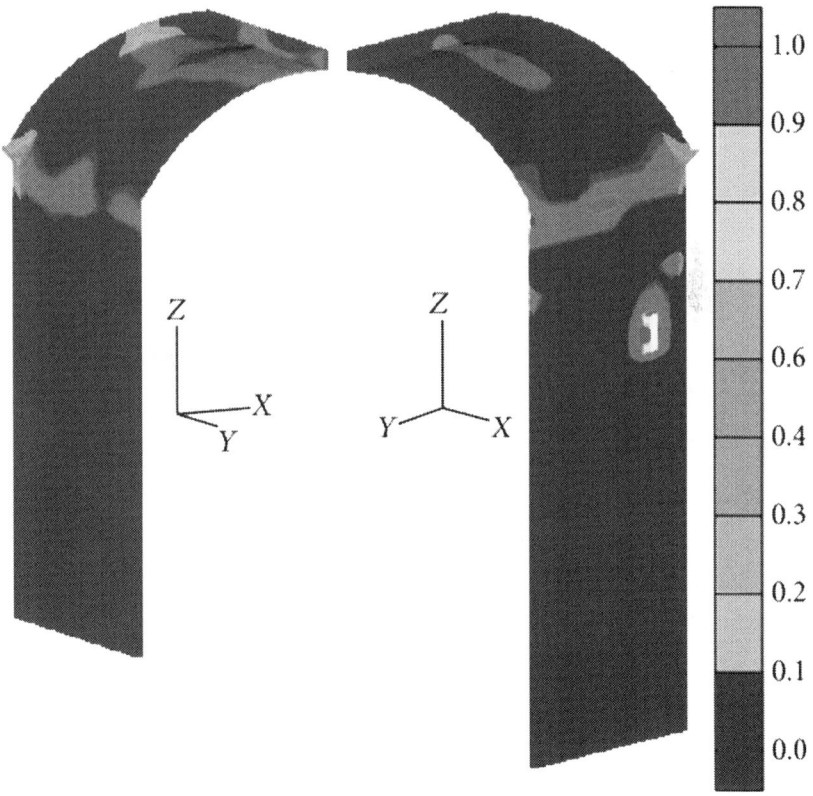

Figure 15: Damage coefficient distribution of the lining when $t = 60$ s.

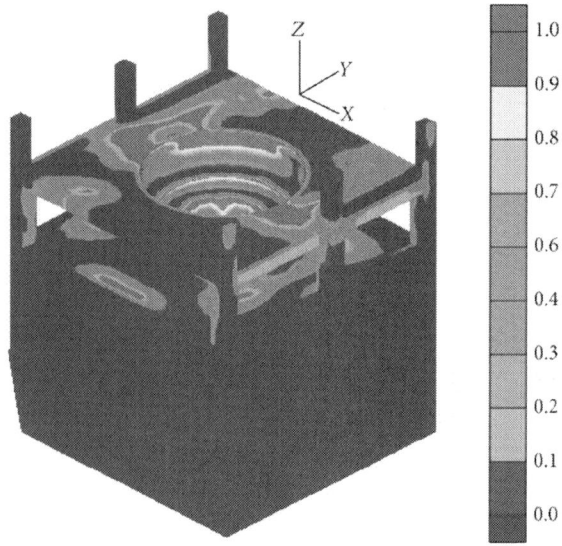

Figure 16: Damage coefficient distribution of the inner concrete structure when $t = 60$ s.

CONCLUSIONS

Based on the dynamic contact force method, which considers the bond-slip properties of the contact surface, a finite element model considering dynamic constitutive properties of materials is built for the dynamic numerical analysis of the underground powerhouse structure of Yingxiuwan Hydropower Plant. The numerical analysis results were compared against the findings from postearthquake investigations. The following conclusions can be made from this study. (1)The dynamic contact force method fully considered the dynamic contact characteristics between the underground powerhouse concrete structure and the surrounding rock. It could simulate three contact states, namely, the cohesive contact, sliding contact, and separation of contact surface. It could be applied effectively in the dynamic calculation of large complex contact systems.(2)The numerical analysis and the postearthquake investigations revealed the evolution process and characteristics of the seismic damage of underground powerhouse structure. The amplitude and duration of seismic waves determined the degree of seismic damage. These are the external factors causing the damage of the underground powerhouse structure. The spatial variations of the structural properties led to the variation in the degree of damage. This is an internal factor causing the damage of the structure. The surrounding rock not only imposed favorable constraint but also caused unfavorable compression on the concrete structure. However, in general, the advantages outweigh the disadvantages. (3) In antiseismic design of the underground powerhouse structure, reinforced support system should be adopted to improve the stiffness of the surrounding rock and reduce the deformation. The upper part of the structure could be strengthened by anchorage support or other support measures to increase the constraining effect of the surrounding rock. The interstitial spaces between the lower parts of the concrete structure and the surrounding rock are recommended to be filled with soft-seismic isolation layer, so as to weaken the unfavorable compression due to surrounding rock and to reduce the seismic damage of the structure.

ACKNOWLEDGMENTS

This study was supported by the National Key Basic Research Program of China (2010CB732005, 2015CB057900), the Major Program of the National Natural Science Foundation of China (91215301), the National Natural Science Foundation of China (51279136, 51209164), and the Research Fund for the Doctoral Program of Higher Education of China (20130141110015). These supports are greatly acknowledged and appreciated.

REFERENCES

1. Z. Z. Shen and H. C. Ren, "Dynamic response characteristics of underground powerhouse caverns for Sandaowan hydropower station," Advanced Materials Research, vol. 382, pp. 80–83, 2012. ·

2. Y. Zhang, M. Xiao, and J. Chen, "Seismic damage analysis of underground caverns subjected to strong earthquake and assessment of post-earthquake reinforcement effect," Disaster Advances, vol. 3, no. 4, pp. 127–132, 2010.

3. B.-Y. Zhao and Z.-Y. Ma, "Influence of cavern spacing on the stability of large cavern groups in a hydraulic power station," International Journal of Rock Mechanics and Mining Sciences, vol. 46, no. 3, pp. 506–513, 2009. · ·

4. W. Q. Sun, Z. Y. Ma, X. Yan, J. Qi, and X. Du, "Intelligent identification of underground powerhouse of pumped-storage power plant," Acta Mechanica Sinica, vol. 21, no. 2, pp. 187–191, 2005. · ·

5. J. Chen, Z. Zhang, and M. Xiao, "Seismic response analysis of surrounding rock of underground powerhouse caverns under obliquely incident seismic waves," Disaster Advances, vol. 5, no. 4, pp. 1160–1166, 2012.

6. O. C. Zienkiewicz, R. L. Taylor, and J. Z. Zhu, The Finite Element Method: Its Basis and Fundamentals, Butterworth-Heinemann, 2005.

7. G. González, M. Gerbault, J. Martinod et al., "Crack formation on top of propagating reverse faults of the Chuculay Fault System, Northern Chile: insights from field data and numerical modelling," Journal of Structural Geology, vol. 30, no. 6, pp. 791–808, 2008. · ·

8. G. E. Hilley, I. Mynatt, and D. D. Pollard, "Structural geometry of Raplee Ridge monocline and thrust fault imaged using inverse Boundary Element Modeling and ALSM data," Journal of Structural Geology, vol. 32, no. 1, pp. 45–58, 2010. · ·

9. D. O. Potyondy and P. A. Cundall, "A bonded-particle model for rock," International Journal of Rock Mechanics and Mining Sciences, vol. 41, no. 8, pp. 1329–1364, 2004. · ·

10. M. Xia and K.-P. Zhou, "Particle simulation of the failure process of brittle rock under triaxial compression," International Journal of Minerals, Metallurgy and Materials, vol. 17, no. 5, pp. 507–513, 2010. · ·

11. M. Xia and C. B. Zhao, "Simulation of rock deformation and mechanical characteristics using clump parallel-bond models," Journal of Central South University, vol. 21, no. 7, pp. 2885–2893, 2014.

12. M. Xia, C. B. Zhao, and B. E. Hobbs, "Particle simulation of thermally-induced rock damage with consideration of

temperature-dependent elastic modulus and strength," Computers and Geotechnics, vol. 55, pp. 461–473, 2014.

13. J. S. Yoon, A. Zang, and O. Stephansson, "Simulating fracture and friction of Aue granite under confined asymmetric compressive test using clumped particle model," International Journal of Rock Mechanics and Mining Sciences, vol. 49, pp. 68–83, 2012. · ·

14. Z. Zhao, L. Jing, and I. Neretnieks, "Particle mechanics model for the effects of shear on solute retardation coefficient in rock fractures," International Journal of Rock Mechanics and Mining Sciences, vol. 52, pp. 92–102, 2012. · ·

15. P. A. Cundall and O. D. L. Strack, "A discrete numerical model for granular assemblies," Geotechnique, vol. 29, no. 1, pp. 47–65, 1979. · ·

16. Z. H. Zhao, "Gouge particle evolution in a rock fracture undergoing shear: a microscopic DEM study,"Rock Mechanics and Rock Engineering, vol. 46, no. 6, pp. 1461–1479, 2013. · ·

17. P. Mora and D. Place, "A lattice solid model for the non-linear dynamics of earthquakes," International Journal of Modern Physics, vol. 6, pp. 1059–1074, 1993.

18. F. Radjaï and F. Dubois, Discrete-Element Modeling of Granular Materials, Wiley-I STE, New York, NY, USA, 2011.

19. L.-J. Dong and X.-B. Li, "Three-dimensional analytical solution of acoustic emission or microseismic source location under cube monitoring network," Transactions of Nonferrous Metals Society of China, vol. 22, no. 12, pp. 3087–3094, 2012. · ·

20. L. J. Dong and X. B. Li, "A microseismic/acoustic emission source location method using arrival times of PS waves for unknown velocity system," International Journal of Distributed Sensor Networks, vol. 2013, Article ID 307489, 8 pages, 2013. ·

21. L. J. Dong, X. B. Li, and G. Xie, "An analytical solution for acoustic emission source location for known P wave velocity system," Mathematical Problems in Engineering, vol. 2014, Article ID 290686, 6 pages, 2014. ·

22. X. B. Li and L. J. Dong, "An efficient closed-form solution for acoustic emission source location in three-dimensional structures," AIP Advances, vol. 4, no. 2, Article ID 027110, 9 pages, 2014.

23. D. M. Potts and L. Zdravkovic, Finite Element Analysis in Geotechnical Engineering, Thomas Telford, 1999.

24. M. Sharafisafa and M. Nazem, "Application of the distinct element method and the extended finite element method in modelling cracks and coalescence in brittle materials," Computational Materials Science, vol. 91, pp. 102–121, 2014. ·

25. G. G. Gray, J. K. Morgan, and P. F. Sanz, "Overview of continuum and particle dynamics methods for mechanical modeling of contractional geologic structures," Journal of Structural Geology, vol. 59, pp. 19–36, 2014.

26. N. Hu, "A solution method for dynamic contact problems," Computers and Structures, vol. 63, no. 6, pp. 1053–1063, 1997. ·

27. D. Peric and D. R. J. Owen, "Computational model for 3-D contact problems with friction based on the penalty method," International Journal for Numerical Methods in Engineering, vol. 35, pp. 1289–1309, 1992.

28. Y. Kanto and G. Yagawa, "Dynamic contact buckling analysis by the penalty finite element method," International Journal for Numerical Methods in Engineering, vol. 29, no. 4, pp. 755–774, 1990. · ·

29. J. C. Simo and T. A. Laursen, "An augmented Lagrangian treatment of contact problems involving friction," Computers & Structures, vol. 42, no. 1, pp. 97–116, 1992. ·

30. T. A. Laursen and V. Chawla, "Design of energy conserving algorithms for frictionless dynamic contact problems," International Journal for Numerical Methods in Engineering, vol. 40, no. 5, pp. 863–886, 1997. ·

31. G. D. Pollock and A. K. Noor, "Sensitivity analysis of the contact/impact response of composite structures," Computers and Structures, vol. 61, no. 2, pp. 251–269, 1996.

32. T. Belytschko, W. K. Liu, and B. Moran, Nonlinear Finite Elements for Continua and Structures, John Wiley & Sons, Chichester, UK, 2000.

33. J. Liu and S. K. Sharan, "Analysis of dynamic contact of cracks in viscoelastic media," Computer Methods in Applied Mechanics and Engineering, vol. 121, no. 1–4, pp. 187–200, 1995. · ·

34. J. Liu, S. Liu, and X. Du, "A method for the analysis of dynamic response of structure containing non-smooth contactable interfaces," Acta Mechanica Sinica, vol. 15, no. 1, pp. 63–72, 1999. · ·

35. J. Lubliner, J. Oliver, S. Oller, and E. Oñate, "A plastic-damage model for concrete," International Journal of Solids and Structures, vol. 25, no. 3, pp. 299–326, 1989. ·

36. J. Lee and G. L. Fenves, "Plastic-damage model for cyclic loading of concrete structures," Journal of Engineering Mechanics, vol. 124, no. 8, pp. 892–900, 1998.

37. J. Lee and G. L. Fenves, "A plastic-damage concrete model for earthquake analysis of dams," Earthquake Engineering & Structural Dynamics, vol. 27, pp. 937–956, 1998.

38. O. Omidi and V. Lotfi, "Finite element analysis of concrete structures using plastic-damage model in 3-d implementation," International Journal of Civil Engineering, vol. 8, no. 3, pp. 187–203, 2010.

39. M. Xiao, "3-D elastoplastic FEM analysis of implicit cylindric anchor bar element for underground opening," Chinese Journal of Geotechnical Engineering, vol. 14, no. 5, pp. 19–26, 1992.

40. Z. G. Zhang, M. X. Xiao, and J. T. C. Chen, "Simulation of earthquake disaster process of large-scale underground caverns using three-dimensional dynamic finite element method," Chinese Journal of Rock Mechanics and Engineering, vol. 30, no. 3, pp. 509–523, 2011.

41. Z. Zhang, J. Chen, and M. Xiao, "Artificial boundary setting for dynamic time-history analysis of deep buried underground caverns in earthquake di saster," Disaster Advances, vol. 5, no. 4, pp. 1136–1142, 2012.

Chapter 7

On the Behavior of Fiberglass Epoxy Composites under Low Velocity Impact Loading

Gautam S. Chandekar[1], Bhushan S. Thatte[2], and Ajit D. Kelkar[2]

[1]Department of Mechanical Engineering, Tennessee Technological University, Cookeville, TN 38505, USA

[2]Computational Science and Engineering, North Carolina A&T State University, Greensboro, NC 27411, USA

ABSTRACT

Response of fiberglass epoxy composite laminates under low velocity impact loading is investigated using LS-DYNA®, and the results are compared with experimental analysis performed using an instrumented impact test setup (Instron dynatup 8250). The composite laminates are manufactured using H-VARTM© process with basket weave E-Glass fabrics. Epon 862 is used as a resin system and Epicure-W as a hardening agent. Composite laminates, with 10 layers of fiberglass fabrics, are modeled using 3D solid elements in a mosaic fashion to represent basket weave pattern. Mechanical properties are calculated by using classical micromechanical theory and assigned to the elements using ORTHOTROPIC ELASTIC material model. The damage occurred since increasing impact energy is incorporated using ADVANCED COMPOSITE DAMAGE material model in LS-DYNA®. Good agreements are obtained with the failure damage results in LS-DYNA® and experimental results. Main considerations for comparison are given to the impact load carrying capacity and the amount of impact energy absorbed by the laminates.

INTRODUCTION

Fiber-reinforced composite materials are extensively used in modern aerospace industry because of their low specific weight with high specific modulus. However out of plane loading, such as impact loading, can cause severe drop in load carrying capacity of these laminates. This drop in load carrying capacity is mainly because of the internal damage of matrix or fiber, which in many times is hard to detect just by visual inspection. In the current work a series of experiments were performed to study the effect of increasing impact energy on fiberglass/epoxy composite laminates in terms of their impact load carrying capacity and impact energy absorption. Fiberglass/epoxy composite laminates are particularly considered as they show superior performance under out of plane loading as opposed to carbon/epoxy laminates, which are strong under in-plane loading. In broad sense the study of low-velocity impact loading on a composite material is divided in three categories, (1) experimental study, (2) analytical study, where a failure model of composite material is proposed, and (3) numerical analysis, mainly using finite element analysis.

Numerous experimental research efforts have been carried out to understand the behavior of composites under low velocity impact applications. When the composite laminate is impacted with the foreign object, the impact dynamics in the vicinity of the impact region becomes very complex [1]. Wang et al. [2] in his research paper discussed the low velocity impact properties of the 3D woven basalt/aramid hybrid composites using experimentally collected data. Tan et al. [3] studied the effect of impacting projectiles with different geometries on the high strength filler fabric. Cheeseman and Bogetti [4] studied the factors affecting the impact forces and the strains of ballistic impact on composite laminates. Bazhenov [5] presented the energy dissipation by bulletproof aramid fabric, while Iremonger and Went [6] studied the mechanism of penetration for ballistic impact on composite armors. Hosur et al. [7, 8] and Siow and Shim [9] studied behavior of various types of composite laminates under low velocity impact and compared their results with ultrasonic c-scans of the damaged laminates. Pearson et al. [10] did global-local assessment of woven composites, where global measurements were taken from low-velocity impact experiments and local strain measurements were obtained using optical Fiber Bragg Grating (FBG) sensors and mapped the failure for two- and three-dimensional woven composites.

Many researchers proposed analytical models for progressive damage in fiber-reinforced composite laminates. Parga-Landa and Hernández-Olivares [11] presented an analytical model to study the impact behavior of soft armors. Robbins and Reddy [12] proposed adaptive hierarchical kinematics approach, which uses variable kinematic finite elements (VKFE) for modeling progressive damage. Shokrieh and Lessard [13, 14] proposed and validated a progressive fatigue damage model for composite laminates. Out of several methods explained above, a damage mechanism proposed by Chang and Chang [15, 16] is used in LS-DYNA® simulations. More detail explanation of this damage model is given in Section 4.1.1.

Apart from these experimental and analytical studies, several numerical simulation approaches are also presented. Mikkor et al. [17] presented finite element model to study the impact behavior of preloaded composites panels. McCarthy et al. [18] studied the bird strike on an aircraft wing leading edge made from fiber-metal laminates using novel SPH finite element approach. Meo et al. [19] studied behavior of composite laminates embedded with shape memory alloy using finite element analysis. Donadon et al. [20] in his research publication proposed a 3D micromechanical analytical model for the study of woven hybrid laminates using CLT. Kermanidis et al. [21] presented a finite element approach for studying the bird impact on the horizontal tail of a transportation aircraft. Ji and Kim [22] used direct numerical simulation to simulate low-velocity impact on 3D orthogonal woven composites.

In modeling woven composites under impact, the experimental and the classical mechanics approach seems to be very expensive and highly complicated. Hence one has to depend on the numerical approaches which are fairly accurate, less expensive and less time consuming. In the present study a finite element approach has been used to model the ten-layered composite laminates using model builder VPG and solved using LS-DYNA® solver. The 10 ply E-glass epoxy laminates have been modeled, and their impact behavior is compared with the experimental results for six different energy levels.

MODELING DETAILS

This section provides details of finite element modeling, used to analyze the impact phenomenon. The modeling of the composite laminate is done using a finite element mesh in mosaic fashion as shown in Figure1(b), to mimic the basket weave pattern shown in Figure 1(a). The laminate model was fully developed using the repeating unit. The repeating unit is the smallest unit that repeats in the whole laminate system in x - y-plane. The fabric consists of fibers oriented in two directions, the warp (0°) and the weft (90°). In the model, the warp and the weft of the composite are represented by a unit cell. The unit cell is created using an 8-node solid element. In the unit cell, the warp and the weft are assigned with orthotropic elastic material properties when "no failure" criteria is used. When composite damage model is used to invoke "failure" criteria, additional strength properties are assigned as shown in Table 2. The resultant properties of the composites unit cell are computed using constituents properties, that is, fiber properties and matrix properties together as shown in Table 1. The unit cell consists of unidirectional fibers impregnated in the epoxy resin, which makes it behave as an orthotropic elastic material making it much stronger in one direction (0°) and weaker in the other (90°) direction.

Table 1: Constituent's elastic properties

Property	Value
Fiber volume fraction Vf	0.50
Matrix volume fraction Vm	0.50

Matrix elastic modulus Em (GPa)	3.45
Matrix shear modulus Gm (GPa)	1.31
Longitudinal elastic modulus Efaa (GPa)	73.09
Transverse elastic modulus Efbb (GPa)	73.09
Shear modulus Gfab (GPa)	30.13
Shear modulus Gfbc (GPa)	30.13

Table 2: Material properties of a composite laminate

Orthotropic elastic material (no failure)		
Property	Warf (0°)	Weft (90°)
Young's modulus a-dir (Ea) (GPa)	38.30	10.56
Young's modulus b-dir (Eb) (GPa)	10.56	38.30
Young's modulus c-dir (Ec) (GPa)	10.56	10.56
Shear modulus (Gab) (GPa)	3.96	3.96
Shear modulus (Gbc) (GPa)	2.45	3.96
Shear modulus (Gca) (GPa)	3.96	2.45
Poisson's ratio (Prba)	0.0787	0.285
Poisson's ratio (Prca)	0.0787	0.4206
Poisson's ratio (Prcb)	0.4206	0.0787
Density (kg/m3)	11.74	
Enhanced composite damage material (with failure)		
Property	Warf (0°)	Weft (90°)
Shear strength in AB plane (SC) (GPa)	8.89E − 02	8.89E − 02
Longitudinal tensile strength (XT) (GPa)	1.08E + 00	1.08E + 00
Transverse tensile strength (YT) (GPa)	3.93E − 02	3.93E − 02
Transverse Compr. strength YC () (GPa)	1.28E− 01	1.28E − 01
Shear stress parameter (ALPH) (GPa)	0.00E + 00	0.00E + 00
Normal tensile strength (SN) (GPa)	6.89E + 09	6.89E + 09
Transverse shear strength (SYZ) (GPa)	6.89E + 09	6.89E + 09
Transverse shear strength (SZX) (GPa)	6.89E + 09	6.89E + 09
Material axes option (AOPT)	0	0

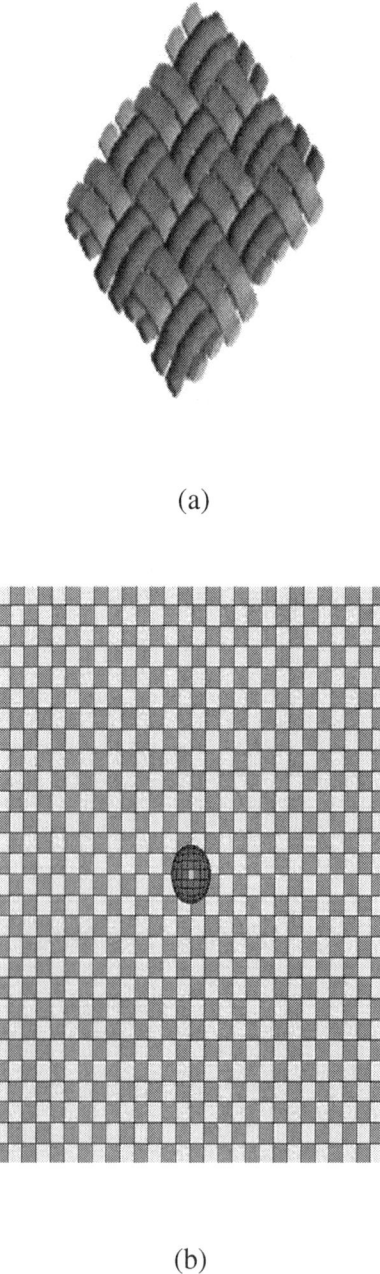

(a)

(b)

Figure 1: Basket weave pattern of Fiberglass/Epoxy laminate and corresponding finite element Mesh.

There are many possible ways to compute the unit cell properties, some based on experimental study and other on parametric approaches. The analytical method of computing properties is time consuming, and the parametric approach requires extensive combinations of tests. Chamis [23] in his research paper presents the simplified equations to compute strengths, fracture toughness, impact properties based on the micromechanics approach. The same equations were adopted for computing the resultant properties of the fiberglass/epoxy unit cells.

Total thickness of the fabricated laminates is measured and accordingly assigned to the elements. The layers are stacked over each other, which are in an alternate (0°–90°) fashion to produce 10-layered composite laminates. Care is taken such that one layer of actual weave is equivalent to two layers in the mosaic FE model. Laminates in the impact test machine holding fixture are clamped at all the four sides. To mimic this boundary condition in modeling, all the boundary nodes are fixed in all translations. Also all the boundary nodes are fixed in all rotations. Solid section is assigned to both warf and weft to simulate the solid structure of the laminate.

The ball is modeled to simulate the 12.7 mm (dia) solid steel impactor of the impact machine. Four-nodded solid quad elements with higher mesh density are assigned to ensure that the cell size of ball is smaller than the cell size of either composite laminate as shown in Figure 2. The steel ball is assigned with solid section similar to the composite laminate. Impactor ball is constrained in 5 degrees of freedom (xand y translation and 3 rotation) allowing its movement only in z direction. The effective density equal to 7860 (kg/m³), Young's modulus equal to 200 GPa, and Poisson's ratio equal to 0.3 are assigned to the material of the ball.

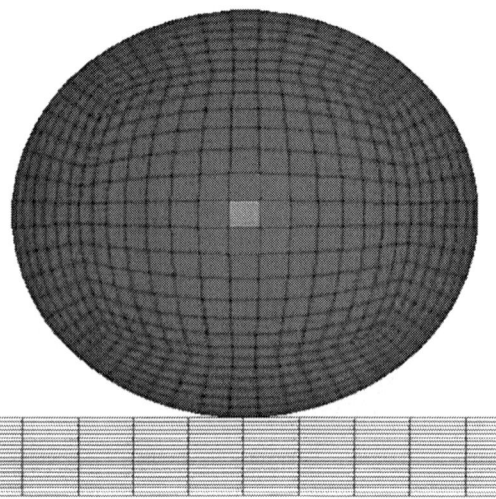

Figure 2: Finite element mesh of an impactor.

In the process of ball impacting the composite plate, the impact starts with the ball touching the composite plate (point to surface contact) and progresses as a surface-to-surface contact. In LSDYNA® an automatic surface-to-surface contact has been assigned between warf and ball and weft and ball to accommodate the impact initiation and the impact progress. The impact time duration, that is, the termination time, computation time step, and output time step are specified in the control cards of LS-DYNA® to control the computational run and to get the output at desired time interval. The output is in the form of binary data, which can be analyzed using the LS-PREPOST postprocessor.

EXPERIMENTAL DETAILS

This section presents detailed explanation of the experimental study conducted.

Fabrication Using H-VARTM© Process

Basket weave woven-roving FGI 1854 E-glass fabrics are manufactured by Fiber Glass Industries, Inc. with EPIKOTE Resin 862 (Bisphenol-F (BPF) epoxy resin) and EPIKURE Curing Agent W (Non-MDA, aromatic amine curing agent) are used to fabricate all the laminates. The fabrication of the composite panels is done using the H-VARTM© (Heated Vacuum Assisted Resin Transfer Molding [25]) process. Four square test coupons of size 0.152 m (6 in) are cut from each panel, which are clamped in impact test machine. The clamped area is 12.7 mm (0.5 in) from all the sides. The total impacted area is $0.127 \times 0.127 m^2$. Figure 3 shows schematic of H-VARTM© setup.

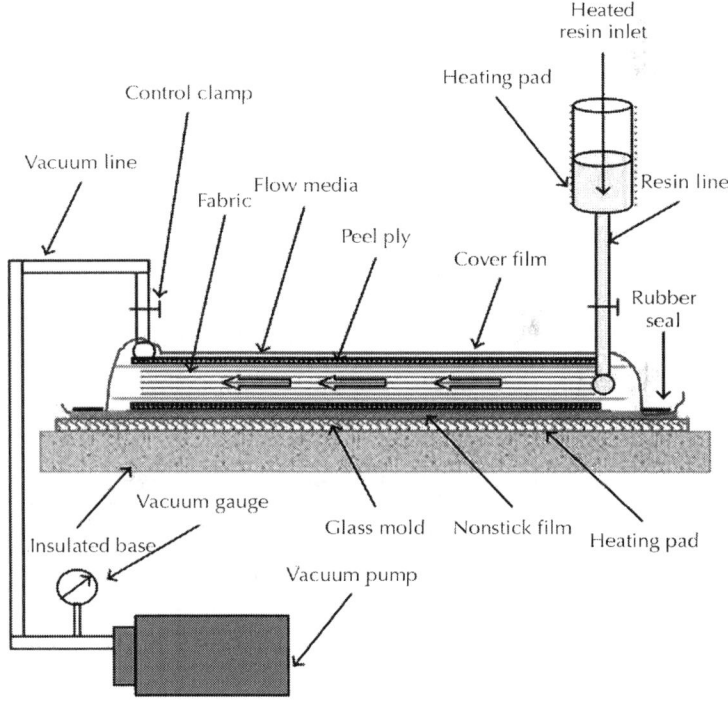

Figure 3: Schematic of H-VARTM© process.

Impact Test Procedure

All the impact tests are performed using the DYNATUP 8250 impact drop tower device as shown in Figure 4. The low velocity impact test facility consists of a drop tower equipped with an impactor and a variable cross-head weight arrangement, a high-speed data acquisition system, and a load transducer mounted in the impactor. In this study the gravity mode is used for all low velocity impact tests. The cross-head/ impactor weight is kept constant for all tests. The cross-head weight is maintained at 5 kg, and including the weight of the impactor, which is 0.45 kg, the total weight becomes 5.45 kg. The low velocity impact facility is equipped with instrumentation to measure the velocity prior to impact. The high-speed data acquisition system has the capability of storing the entire impact event and produce load-time, load-deflection, and energy-time curves. The objectives of the preliminary impact tests are

- (to establish the energy levels and drop height for the incipient damage threshold or lower bound,
- to establish the energy levels and drop height for penetration or the upper bound.

(a)

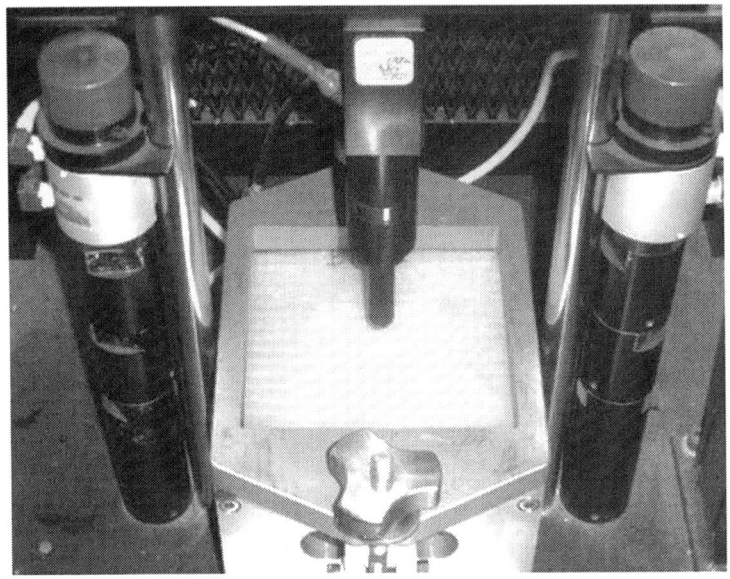

(b)

Figure 4: DYNATUP 8250 test setup.

To achieve the above objectives, a series of impact tests are performed and are discussed in the following paragraph. To start with, a random drop height is selected to perform the low velocity impact test on the woven composites. After impact, the specimens are examined for damage. The impact height is varied until the impact load-time history plots indicated no impact damage. The energy level corresponding to this drop height is called threshold energy level or lower bound. To establish the energy level for the upper bound (i.e., beginning of penetration), the drop height is selected such that there will be a decrease in impact load carrying capacity of the laminate if this height is increased further. Once the lower and upper bound energy levels were established, the difference in the drop heights was calculated. The loading range is divided into six (6) successive drop heights, and the three impact tests are carried out for each drop height to obtain the statistical accuracy. This series of tests gives sufficient data to analyze the damage characteristics and to study the progression of damage in fiberglass/epoxy composite laminates.

RESULTS AND DISCUSSIONS

This section present a discussion of (a) comparative results obtained from experimental study and LS-DYNA® simulations and (b) progressive damage.

Comparative Results

Comparative plots of experimental and LS-DYNA® results for E-Glass/Epoxy laminates under six different impact energy levels are reported in this section. LS-DYNA® runs are conducted with and without the failure criteria and compared with experimental results. The lowest energy level is chosen from the incipient damage seen in the test, and the impact energy is increased with equal increments to incorporate six energy levels. Table 3 presents experimental results for six drop heights (impact energy levels) of an impactor with three repetitions per drop height. Average value of three runs for each drop height is considered in all the comparison plots. As seen in Table 3, the maximum load carrying capacity of E-glass laminates increases with the increase in impact energy. This indicates that the maximum load carrying capacity (max. impact load) of the laminates has not yet reached to the saturation level, where considerable damage is seen.

Table 3: E-Glass epoxy experimental results

No.	Impact level	Drop height (m)	Laminate thickness (mm)	Max impact load (N)	Time to reach max load (ms)	Impact velocity (m/s)	Impact energy (J)	Impact duration (ms)
1	1	0.10	5.11	2770.88	3.64	1.30	0.72	7.77
2	1	0.10	5.31	2865.31	3.65	1.36	1.19	7.79
3	1	0.10	4.90	3069.12	3.47	1.38	0.93	7.34
4	2	0.20	5.33	4323.90	3.52	1.92	0.93	7.27
5	2	0.20	5.21	3965.93	3.58	1.87	1.46	7.85
6	2	0.20	4.90	4495.28	3.12	1.90	1.26	6.93
7	3	0.30	4.80	4927.67	4.02	2.37	2.59	8.13
8	3	0.30	4.75	4768.39	4.02	2.37	3.74	8.34
9	3	0.30	4.90	4700.65	4.26	2.38	4.08	8.50

10	4	0.41	4.98	6094.38	3.72	2.74	2.76	7.66
11	4	0.41	4.75	6015.70	3.56	2.76	4.85	7.47
12	4	0.41	4.83	5930.16	3.72	2.76	5.03	7.55
13	5	0.51	4.60	7364.69	3.46	3.10	3.54	7.08
14	5	0.51	4.50	6822.34	3.80	3.10	6.54	8.02
15	5	0.51	4.62	6896.76	3.91	3.13	6.12	8.01
16	6	0.61	4.57	8122.94	3.62	3.42	4.30	7.48
17	6	0.61	4.62	7718.70	3.78	3.41	9.21	8.06
18	6	0.61	4.95	7879.41	3.76	3.42	7.36	7.89

To understand the behavior of E-glass/epoxy laminates with respect to time, comparison plots of "impact load" and "impact energy loss" versus time are given in Figures 5 and 6, respectively. As seen in Figure 5 when "no failure" criteria is used, the material overpredicts maximum load carrying capacity, and the impact time is less than the experimental values. Since elastic material is used and no damage mechanism is incorporated in "no failure" LS-DYNA® runs, it justifies the increasing trend of impact load with overpredicted values. The results indicate that the material with "no failure" criteria is slightly stiffer than the actual material. When the "failure" criteria is invoked, it can be seen that the loading versus time plot is in better agreement with experimental results.

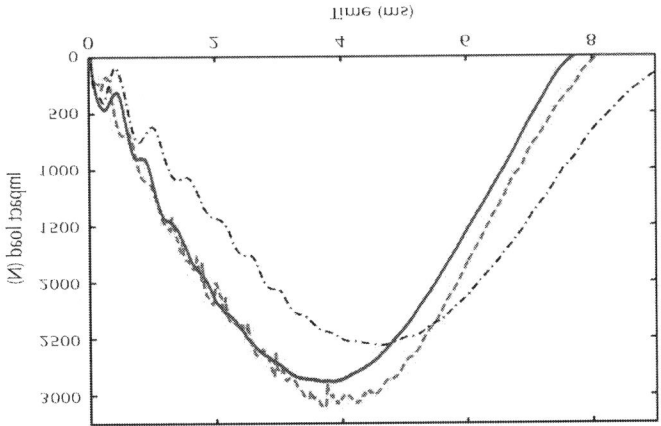

(a)

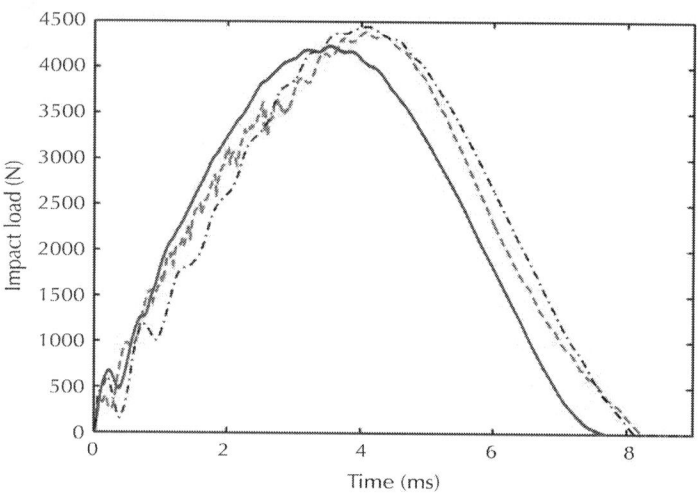

(b)

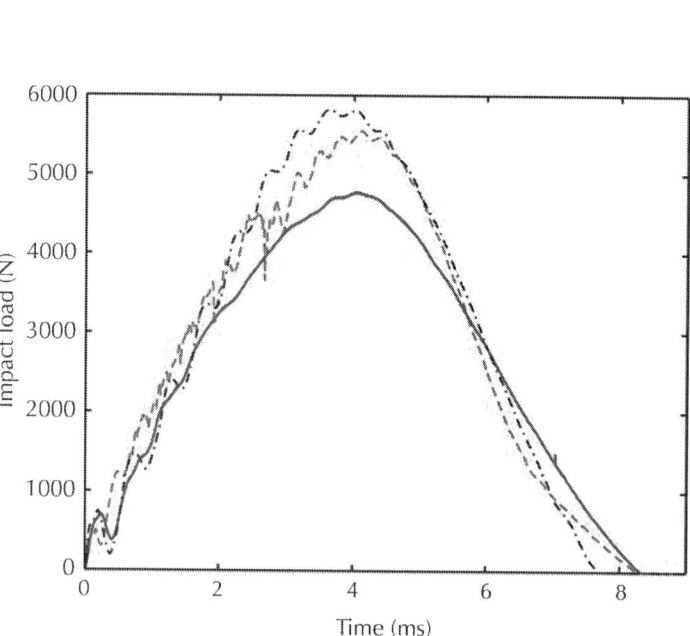

(c)

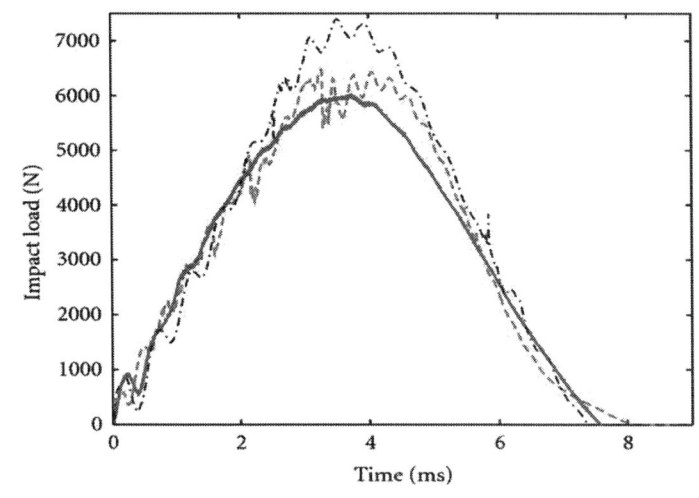

(d)

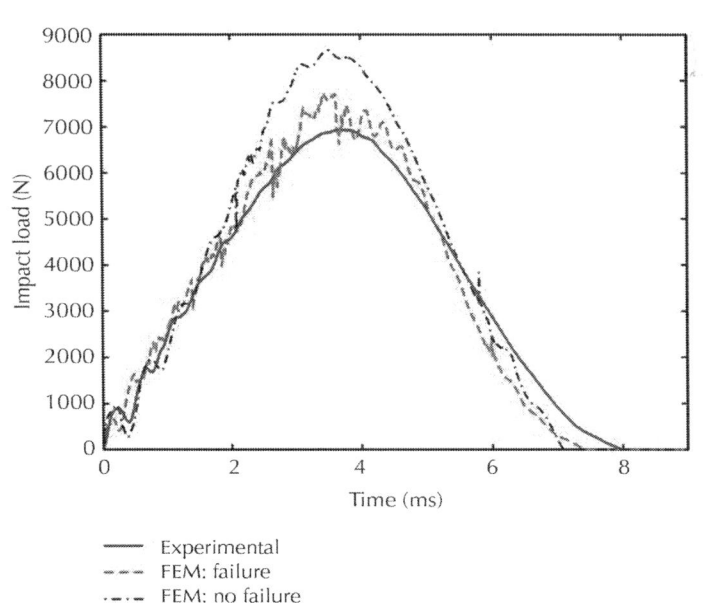

(e)

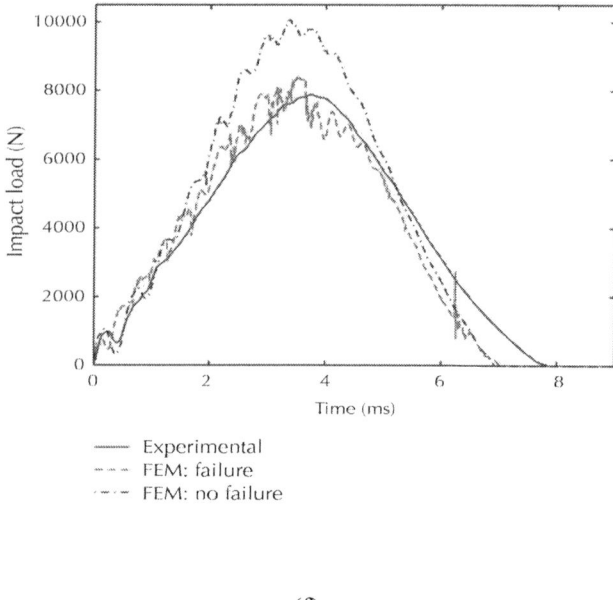

(f)

Figure 5: Impact load versus time plots for E-Glass/epoxy laminates at six impact energy levels.

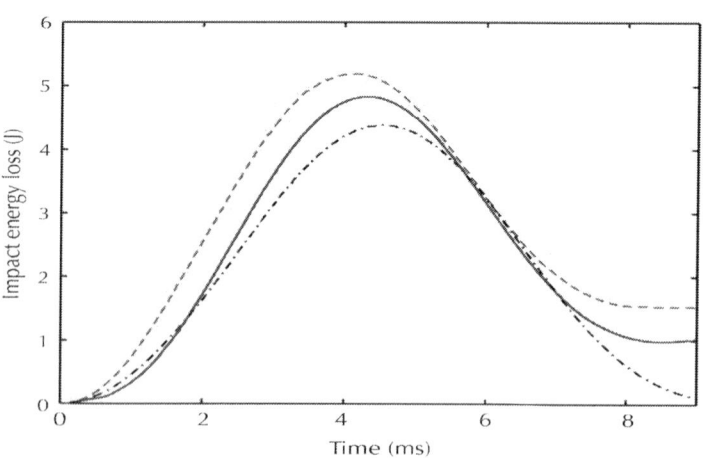

(a)

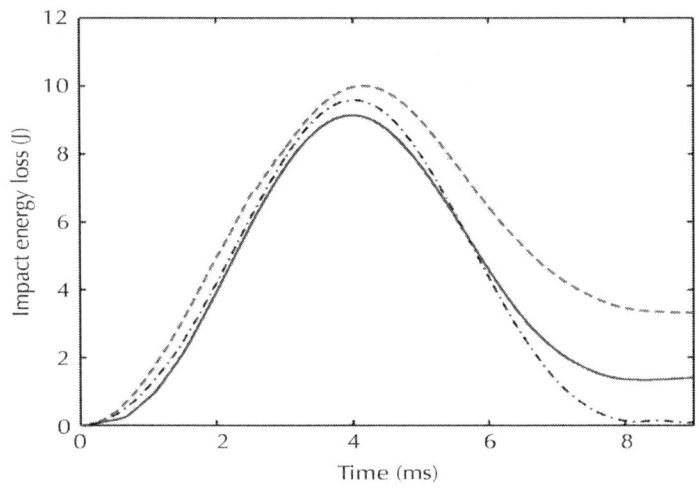

(b)

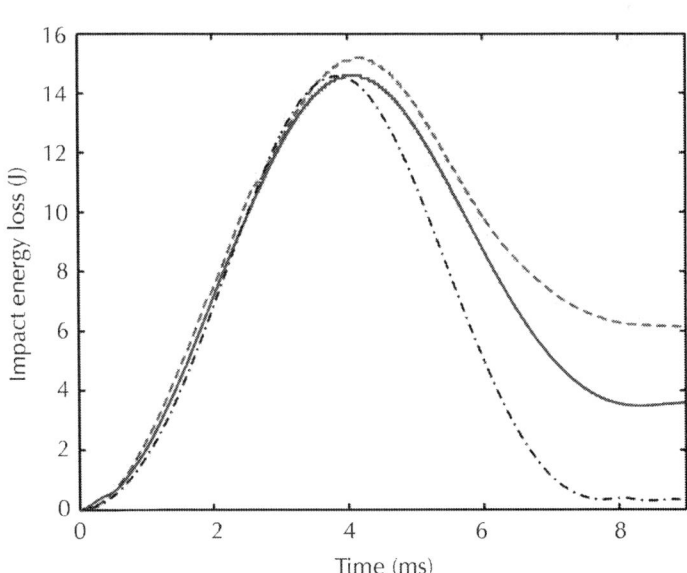

(c)

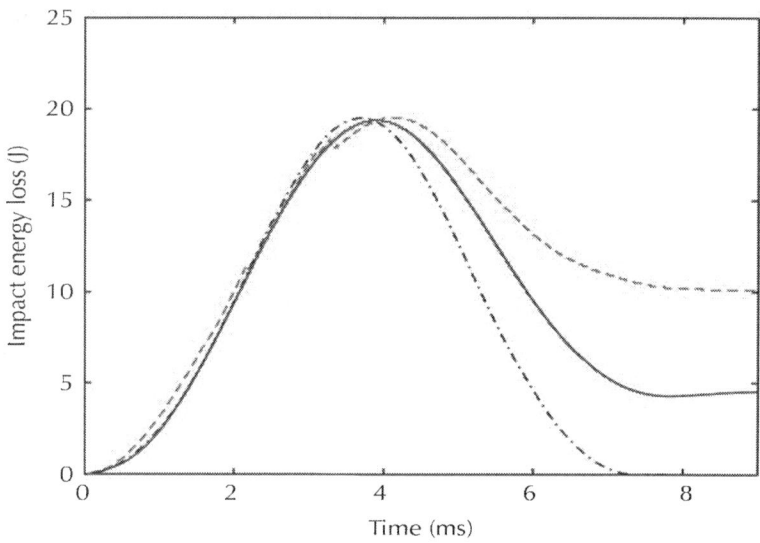

(d)

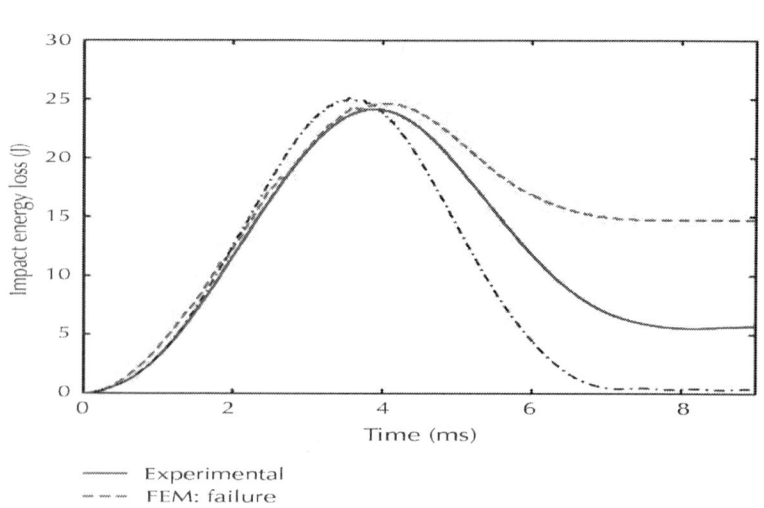

(e)

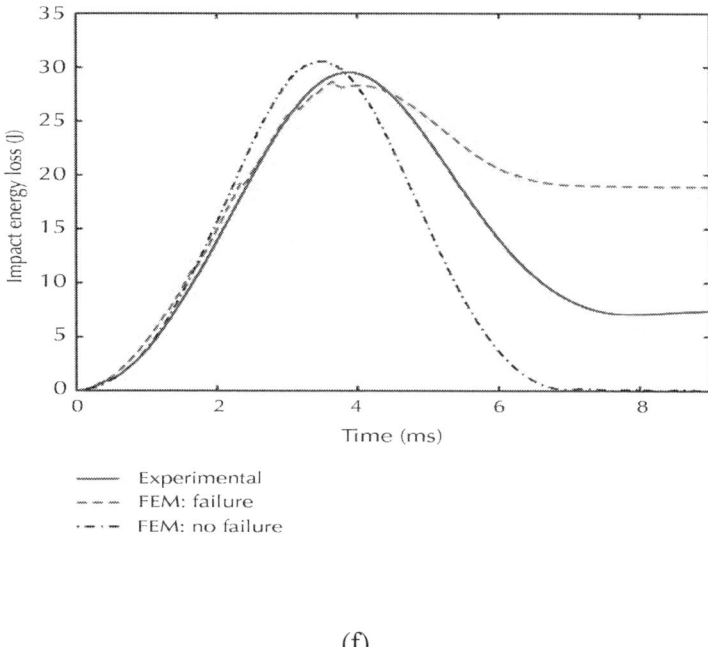

Experimental
FEM: failure
FEM: no failure

(f)

Figure 6: Impact energy loss versus time plots for E-Glass/epoxy laminates at six impact energy levels.

To get more insight of composite laminate behavior, impact energy loss versus time plots for six energy levels are shown in Figure 6. When "no failure" criteria is used, that is, for orthotropic elastic material, no energy is absorbed by the laminate during the impact. Consequently the energy curve reaches to zero at the end of impact showing material's elastic deformation. In experimental case and where "failure" criteria is invoked, the deformation is not elastic because of the damage and impact energy loss is not zero. However, when "failure" criteria is used, the actual absorbed energy is overpredicted. The subsequent section (Section 4.1.1) presents detail discussion of failure criteria invoked in LS-DYNA®.

Invoking Failure Criteria

MAT_COMPOSITE_DAMAGE material is invoked in LS-DYNA®, which primarily uses F. Chang and K. Chang [15, 16] Composite Failure Model

(Material model 22.1 in LS-DYNA®). The model is based on following five material properties used in three failure modes.

- S1, Longitudinal Tensile Strength.
- S2, Transverse Tensile Strength.
- S12, Shear Strength.
- C2, Transverse Compressive Strength.
- α, Nonlinear Shear Stress Parameter.

In its simplest form the model represents the orthotropic material failure without considering the interlaminar delamination. The model does have a provision to incorporate laminated shell theory to invoke the transverse shear deformation, which is insignificant for low velocity impact problem because primarily failure occurs because of the in-plane stresses in the laminate and small amount of crushing stresses.

The in-plane failure is often the dominant mechanism in tensile failure of fiber-dominated laminated composites. The three different in-plane failure modes are given as matrix cracking, fiber-matrix shearing, and fiber breakage. When the certain element in the FE model is failed as mentioned in any of the failure modes, certain parameters are set to zero. Consequently the load carrying capacity of that element reduces to zero.

In the present problem of low velocity impact the composite damage can be addressed as a combination of matrix failure and fiber failure. Actual failure of the laminate initiates when the first matrix cracking occurs. It then continues when the first fiber breaks and when substantial number of fibers break the element which in turn loses its load carrying capacity completely, which we call the failure of the laminate. The complete span between damage initiations till complete failure of the laminate is very complex to model because at every loading stage the load is shared by matrix and fibers. Also at every failing stage an interlaminar failure occurs along with the progressive damage of the fibers impregnated in the matrix resin. In the mosaic model when the failure stress is reached at any point in the element, the element completely looses its load carrying capacity, and further load is shared by surrounding elements eventually causing them to fail.

In order to bridge the gap between the simple orthotropic material failure model, that is, Chang-Chang Composite Failure Model and the interlaminar progressive damage material model, it is necessary to counterbalance the loss in strength by making failure criteria much stiffer to mimic the actual failure of the laminate which can then be extended over the complete loading range. After numerous iterations, we found that the Enhanced Composite Damage Model (Material model 22 in LS-DYNA®) gives superior results for mosaic model.

The mosaic model in LS-DYNA® delineates soft impact response for higher impact energy levels because of the above-mentioned reasons. The impact load-time response seems to be in good agreement with the experimental response. However because of the successive damage in the elements the energy loss in the simulated model increases as the impact energy level increases.

Progressive Damage

Figure 7 shows the progressive damage due to increasing impact energy levels. From the experimental impacted specimen, it is observed that the damage is prominent on the back face of the laminate, and its negligible on the impacted surface over the complete loading range. This is because the top surface of the laminate undergoes compressive stresses and crushing stresses whereas the bottom surface undergoes tensile stresses. The fiber breaking mainly occurs due to the tensile stresses. It is also seen that the maximum damage zone can be seen on the back face of the laminate. The experimental damage zone is compared with the simulated damage zone (plastic strain contour), and it can be seen that the simulated damaged region is in good agreement with the experimental damaged region. In the glass laminates which are semitranslucent, the damage can be easily seen by naked eyes unlike carbon laminates where ultrasonic c-scan will be required to detect the damage area. Figure 7 shows the comparison of the experimental damage area and the simulated damage area. The fringes in the simulated model show the plastic strain contours.

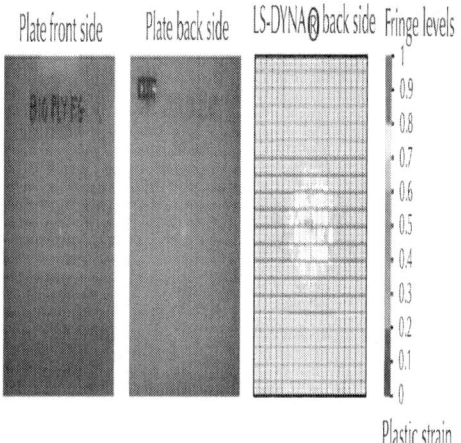

(a)

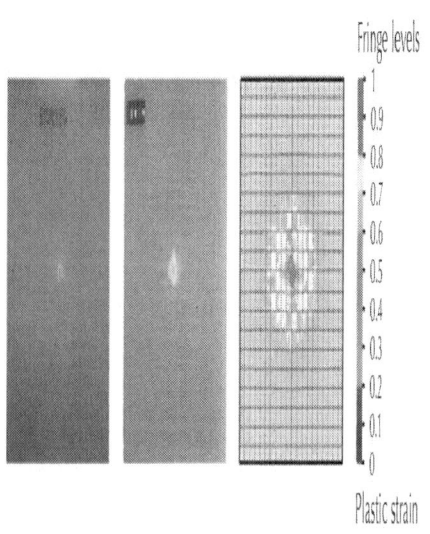

(b)

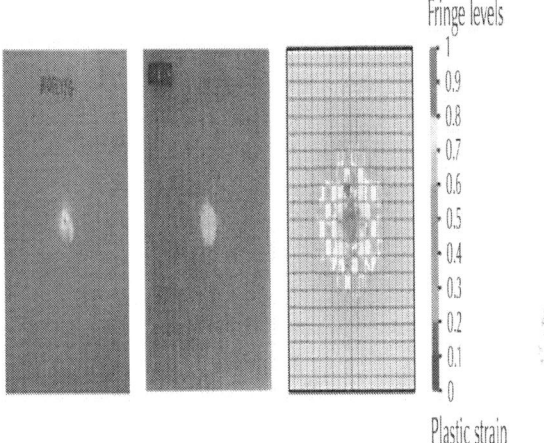

(c)

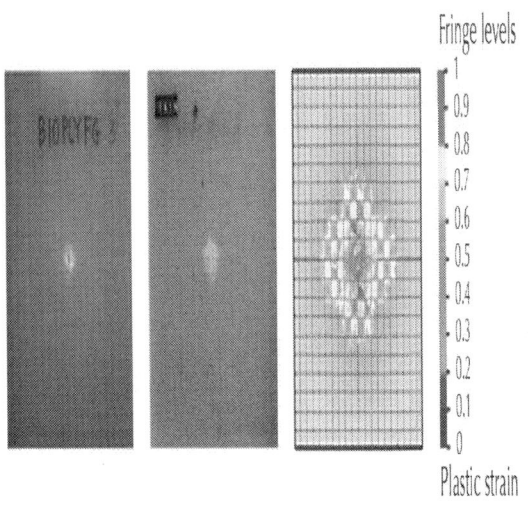

(d)

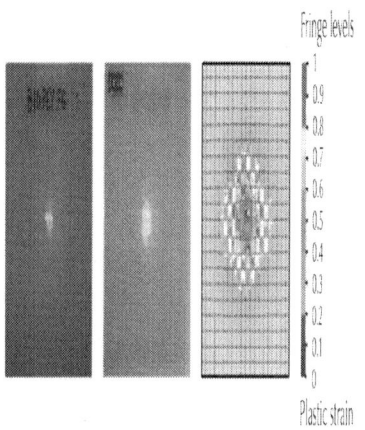

(e)

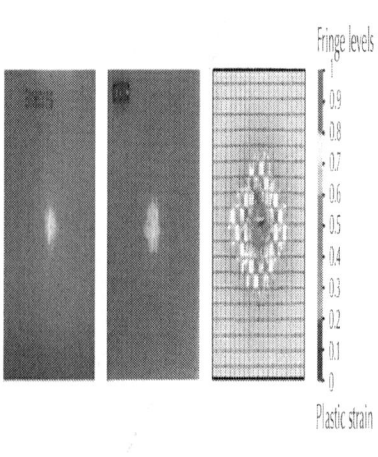

(f)

Figure 7: Progressive damage in E-Glass laminate at 6 impact energy levels (a) 5.45 J, (b) 10.9 J, (c) 16.35 J, (d) 21.8 J, (e) 27.25 J, and (f) 32.7 J.

CONCLUSIONS

In the present study, response of fiberglass/epoxy composite laminates under low velocity impact loading is investigated using LS-DYNA®. The composite laminates are manufactured by the H-VARTM© process using basket weave E-Glass fabrics with the Epon 862 resin system and Epicure-W as a hardening agent. The LS-DYNA® runs are performed using orthotropic elastic material with "no failure" criteria as well as with damage material with "failure" criteria. These results are then compared with experimental drop test results. The experiments are conducted using Instron Dynatup Low Velocity Impact Test Machine. The "no failure" material model in LS-DYNA® overpredicted the load carrying capacity and showed no energy absorption after impact. These results are justified considering its pure elastic behavior, where total energy is regained after impact is complete. However, the "failure" material underpredicted the load carrying capacity and overpredicted amount of impact energy absorbed by laminates. The overprediction of impact energy absorption can be attributed to the basic geometry of the laminate where no undulations between two layers of the laminate are modeled. Since there is only nodal contact contact between warf and weft (mosaic fashion of elements), the total laminate is acting as a softer material than it is in reality. Also, when failure starts to occur, FEA code starts deleting the elements and load carrying capacity decreases. The present study can be easily extended for other weave patterns including twill weave, satin weave, and braided weave composites.

DISCLOSURE

G. Chandekar, author of the paper and all other authors, B. Thatte and A. Kelkar, have no direct financial relation with the commercial software (code) LS-DYNA®. Any other commercial or home grown Finite Element code can be used to verify all the results produced in this paper. G. Chandekar and B. Thatte also have no financial gain by using H-VARTM© process of manufacturing E-glass/Epoxy composite laminates. A patent filed by A. Kelkar and Ronnie Bolick is pending on the H-VARTM© process.

REFERENCES

1. S. Abrate, "Impact on laminated composite materials," Applied Mechanics Reviews, vol. 44, no. 4, p. 155, 1991.

2. X. Wang, B. Hu, Y. Feng et al., "Low velocity impact properties of 3D woven basalt/aramid hybrid composites," Composites Science and Technology, vol. 68, no. 2, pp. 444–450, 2007.

3. V. B. C. Tan, C. T. Lim, and C. H. Cheong, "Perforation of high-strength fabric by projectiles of different geometry," International Journal of Impact Engineering, vol. 28, no. 2, pp. 207–222, 2003.

4. B. A. Cheeseman and T. A. Bogetti, "Ballistic impact into fabric and compliant composite laminates,"Composite Structures, vol. 61, no. 1-2, pp. 161–173, 2003.

5. S. Bazhenov, "Dissipation of energy by bulletproof aramid fabric," Journal of Materials Science, vol. 32, no. 15, pp. 4167–4173, 1997.

6. M. J. Iremonger and A. C. Went, "Ballistic impact of fibre composite armours by fragment-simulating projectiles," Composites Part A: Applied Science and Manufacturing, vol. 27, no. 7, pp. 575–581, 1996.

7. M. V. Hosur, M. Adbullah, and S. Jeelani, "Studies on the low-velocity impact response of woven hybrid composites," Composite Structures, vol. 67, no. 3, pp. 253–262, 2005.

8. M. V. Hosur, F. Chowdhury, and S. Jeelani, "Low-velocity impact response and ultrasonic NDE of woven carbon/epoxy-nanoclay nanocomposites," Journal of Composite Materials, vol. 41, no. 18, pp. 2195–2212, 2007.

9. Y. P. Siow and V. P. W. Shim, "An experimental study of low velocity impact damage in woven fiber composites," Journal of Composite Materials, vol. 32, no. 12, pp. 1178–1202, 1998.

10. J. D. Pearson, M. A. Zikry, M. Prabhugoud, and K. Peters, "Global-local assessment of low-velocity impact damage in woven composites," Journal of Composite Materials, vol. 41, no. 23, pp. 2759–2783, 2007.

11. B. Parga-Landa and F. Hernández-Olivares, "An analytical model to predict impact behaviour of soft armours," International Journal of Impact Engineering, vol. 16, no. 3, pp. 455–466, 1995.

12. D. H. Robbins Jr. and J. N. Reddy, "Adaptive hierarchical kinematics in modeling progressive damage and global failure in fiber-reinforced composite laminates," Journal of Composite Materials, vol. 42, no. 2, pp. 143–172, 2008.

13. M. M. Shokrieh and L. B. Lessard, "Progressive fatigue damage modeling of composite materials, part I: modeling," Journal of Composite Materials, vol. 34, no. 13, pp. 1056–1080, 2000.

14. M. M. Shokrieh and L. B. Lessard, "Progressive fatigue damage modeling of composite materials, part II: material characterization and model verification," Journal of Composite Materials, vol. 34, no. 13, pp. 1081–1116, 2000.

15. F. Chang and K. Chang, "Progressive damage model for laminated composites containing stress concentrations," Journal of Composite Materials, vol. 21, no. 9, pp. 834–855, 1987.

16. F. Chang and K. Chang, "Post-failure analysis of bolted composite joints in tension or shear-out mode failure," Journal of Composite Materials, vol. 21, no. 9, pp. 809–833, 1987

17. K. M. Mikkor, R. S. Thomson, I. Herszberg, T. Weller, and A. P. Mouritz, "Finite element modelling of impact on preloaded composite panels," Composite Structures, vol. 75, no. 1–4, pp. 501–513, 2006.

18. M. A. McCarthy, J. R. Xiao, C. T. McCarthy et al., "Modelling of bird strike on an aircraft wing leading edge made from fibre metal laminates—part 2: modelling of impact with SPH bird model," Applied Composite Materials, vol. 11, no. 5, pp. 317–340, 2004.

19. M. Meo, E. Antonucci, P. Duclaux, and M. Giordano, "Finite element simulation of low velocity impact on shape memory alloy composite plates," Composite Structures, vol. 71, no. 3-4, pp. 337–342, 2005.

20. M. V. Donadon, B. G. Falzon, L. Iannucci, and J. M. Hodgkinson, "A 3-D micromechanical model for predicting the elastic behaviour of woven laminates," Composites Science and Technology, vol. 67, no. 11-12, pp. 2467–2477, 2007.

21. Th. Kermanidis, G. Labeas, M. Sunaric, and L. Ubels, "Development and validation of a novel bird strike resistant composite leading edge structure," Applied Composite Materials, vol. 12, no. 6, pp. 327–353, 2005.

22. K. H. Ji and S. J. Kim, "Dynamic direct numerical simulation of woven composites for low-velocity impact," Journal of Composite Materials, vol. 41, no. 2, pp. 175–200, 2007.

23. C. C. Chamis, "Simplified composite micromechanics equations for strength, fracture toughness and environmental effects," pp. 16–35, 1984.

24. J. Whitcomb and X. Tang, "Effective moduli of woven composites," Journal of Composite Materials, vol. 35, no. 23, pp. 2127–2144, 2001.

25. R. Bolick and A. D. Kelkar, "Innovative composite processing by using H-VARTM© method," inProceedings of the 28th International Conference on SAMPE Europe, 2007.

Failure Mechanism of Rock Bridge Based on Acoustic Emission Technique

Guoqing Chen, Yan Zhang, Runqiu Huang, Fan Guo,
and Guofeng Zhang Kelkar[2]

State Key Laboratory of Geohazard Prevention and Geoenvironment Protection, Chengdu University of Technology, Chengdu, Sichuan 610059, China

ABSTRACT

Acoustic emission (AE) technique is widely used in various fields as a reliable nondestructive examination technology. Two experimental tests were carried out in a rock mechanics laboratory, which include (1) small scale direct shear tests of rock bridge with different lengths and (2) large scale landslide model with locked section. The relationship of AE event count and record time was analyzed during the tests.

The AE source location technology and comparative analysis with its actual failure model were done. It can be found that whether it is small scale test or large scale landslide model test, AE technique accurately located the AE source point, which reflected the failure generation and expansion of internal cracks in rock samples. Large scale landslide model with locked section test showed that rock bridge in rocky slope has typical brittle failure behavior. The two tests based on AE technique well revealed the rock failure mechanism in rocky slope and clarified the cause of high speed and long distance sliding of rocky slope.

INTRODUCTION

Acoustic emission (AE) is a phenomenon about the development of transient elastic waves caused by rapid energy release due to crack growth [1]. Early AE technique has been applied in security problems of metal mines, coal mines, and tunnel engineering. With advanced computerized technique, small sized digital AE analysis equipment can be used to detect the defects and its location and failure propagation. Compared with microseism [2] and other in situ detection methods [3, 4], AE technique can forecast brittle failure before the structural failure.

Rocky landslides contain a lot of rock bridges, and researches on rock bridge were carried out by experimental methods and simulated by analytical and numerical simulation techniques. The laboratory study of AE characteristics during rock stress-strain process was reviewed [5]; real-time failure process [6, 7] and stability evaluation and harm degree of landslides [8] were marked out. Behavior of rock bridge was observed firstly by shear test in 1990 [9]; then cyclic loading test of rock bridge and mechanical behavior was carried out [10]. The failure mechanisms and pattern of coalescence of rock bridges were investigated [11]. Strength, deformability, failure behavior, and AE locations of red sandstone were tested by using triaxial compression [12]. The abovementioned tests showed the basic mechanical behavior, but the brittle failure of rock bridge was rarely tested. Analytical methods such as neural network [13], slice element method [14], and time-dependent degradation failure [15, 16] are used to study the progressive failure of rock bridge. Rock fragmentation processes subjected to static and dynamic loading were examined by the numerical code RFPA [17]. The progressive rock fracture mechanism of cracked chevron notched Brazilian disc rock specimens was numerically simulated [18]. Surface crack initiation and propagation were numerically investigated via parallel finite element analysis [19]. These numerical studies really revealed the progressive failure of rock bridge.

However, the brittle failure of the rock bridge, especially in the rocky slope, is rarely studied. Small scale rock bridge direct shear test and large scale landslide model test with locked section were designed in this paper. The relation curve of AE event count with time and AE source location are obtained through the AE source location technology. The mechanical characteristics and failure mechanism of rock mass with rock bridge and the space evolution process of cracks are discussed.

AE TECHNIQUE

AE Instrument and its Principle

AE instrument is Micro-II Digital AE System, which is equipped with the third generation of digital system developed by USA Physical Acoustics Company. Its core is an AE function card with 8 channels on a plate. Compared to other AE devices, Micro-II Digital AE System is one of the latest and highest integration AE systems currently which has the advantages of small volume, convenient carrying, high precision, high speed of data transmission, strong processing ability, and so on. The data obtained from Micro-II Digital AE System is more reliable and AE source location is more precise than other AE devices. Its maximum signal amplitude reached 100 dB and its bandwidth is 1~400 kHz.

Peripheral equipment mainly includes 2/4/6 type preamplifier with its signal gain ranging from 20 dB, 40 dB, or 60 dB (adjustable), which has a high pass filtering function of 20 Hz. Nano-30 type sensors, whose frequency response range is 100~400 kHz and test set monitoring threshold value is 40 dB, use six sensors array data acquisition to ensure the location precision.

AE principle is that elastic wave from AE source which finally transmits to the material surface causing surface displacement, which can be detected by AE sensors. These sensors convert mechanical vibration into electrical signals. Finally, according to the analysis of the obtained data, the following goals can be achieved: (1) determining the AE source location; (2) determining time or load of the AE occurrence; and (3) assessing the degrees of damage to the AE source.

Figure 1 shows AE signal waveform and partial of AE parameters, including AE count, AE amplitude, rise time, AE duration, and AE threshold. In addition, the commonly used AE characterization parameters also include energy, impact, events, energy, time difference, and frequency [20].

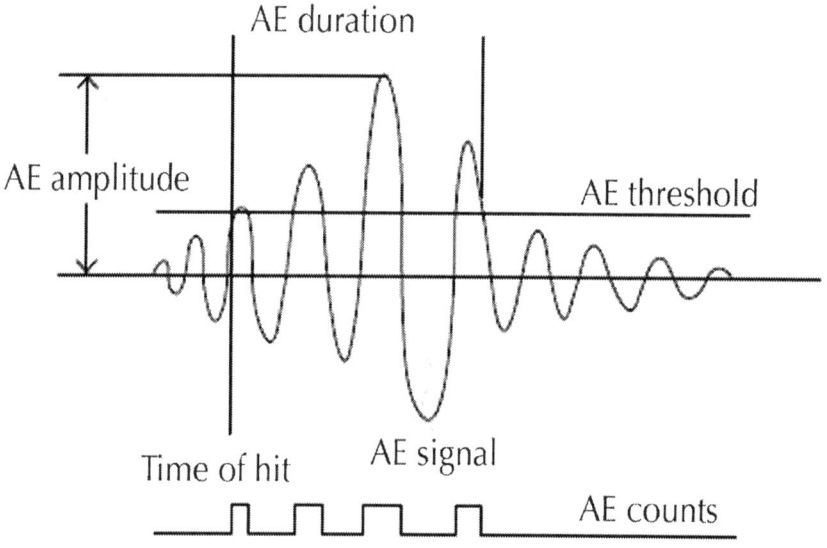

Figure 1: Wave hit feature of AE.

AE Location Technology

AE source location process is shown in Figure 2; different methods of AE source location are adapted for burst and continuous AE signals [21]. Source location of burst AE signal includes time of arrive (TOA) location and region location. Continuous AE signal location includes amplitude measurement type regional location, attenuation measurement type location, cross correlation TOA location, and interference TOA location.

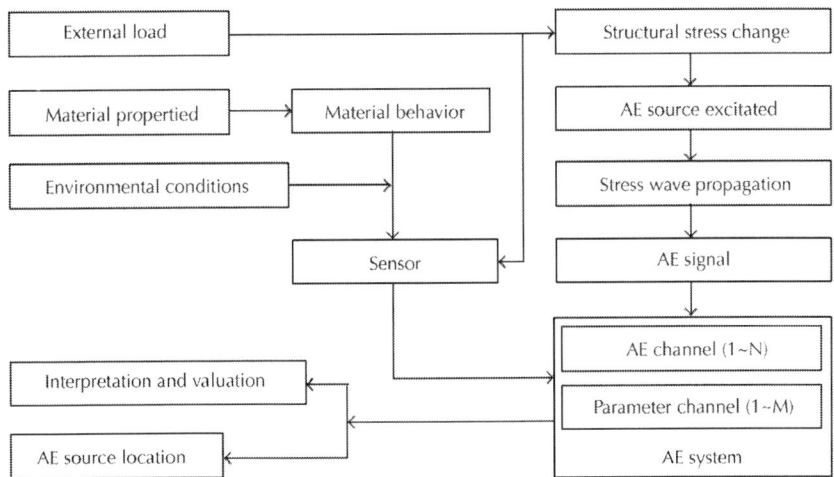

Figure 2: Flow chart of AE source location.

TOA location is widely used for detecting the samples and components, but it would easily lose a large number of low amplitude signals and its location accuracy may be affected by wave speed, attenuation, wave shape, and the shape of component. When the ratio of the length to the radius of the tested object becomes very large, it is appropriate by using linear location for AE detection, such as pipes, bars, and steel beam. Linear location requires at least two AE sensors. Plane location of plate structure needs at least three sensors. Three-dimensional location requires at least four sensors.

The position of sound source can be calculated by the above equation. The basic principle of plane location and three-dimensional location is similar to the linear location, which are all positioned by the velocity and time difference. In this paper, the main AE source location is Geiger method [22] which uses the difference arrive time of P wave [23].

The Geiger method is an application of the Gauss-Newton minimum fitting function; the principle is using iteration method to get the final result from a given initial point (test point). A correction vector $\Delta\theta$ is calculated through every iteration based on the least square method; the vector $\Delta\theta(\Delta x, \Delta y, \Delta z, \Delta t)$ is added to the previous iteration results (test point) to get a new test point. Then it is determined whether the new test point meets the requirements. If it does, the point is the sought source point, if not, the iteration continues until the requirement is met. The result of each iteration is produced by the following time distance equation:

$$\left[(x_i - x)^2 + (y_i - y)^2 + (z_i - z)^2\right]^{1/2} = v_p(t_i - t),$$

(1)

where x,y,z is the test point coordinate (initial value is artificially settled), t is the time of occurrence of events (initial value is artificially settled), x_i, y_i, z_i is the sensor i location, t_i is the arrival time of P wave which is detected by the sensor i, and v_p is the P wave velocity.

t_{oi} is the time of P wave arrival of each sensor, which can be calculated by the First Order Taylor Series Expansion of arrival time through the test point coordinate:

$$t_{oi} = t_{ci} + \frac{\partial t_i}{\partial x}\Delta x + \frac{\partial t_i}{\partial y}\Delta y + \frac{\partial t_i}{\partial z}\Delta z + \frac{\partial t_i}{\partial t}\Delta t,$$

(2)

Where t_{oi} is the arrival time of P wave detected by the sensor i and t_{ci} is the arrival time of sensor i of P wave calculated by test point coordinates.

In (2),

$$\frac{\partial t_i}{\partial x} = \frac{(x_i - x)}{vR}, \qquad \frac{\partial t_i}{\partial y} = \frac{(y_i - y)}{vR}, \qquad \frac{\partial t_i}{\partial z} = \frac{(z_i - z)}{vR},$$

$$\frac{\partial t_i}{\partial t} = 1, \qquad R = \left[(x_i - x)^2 + (y_i - y)^2 + (z_i - z)^2\right]^{1/2}.$$

(3)

For N positive sensors, you can get N equations written in matrix form:

$$A\Delta\theta = B,$$

(4)

where

$$A = \begin{bmatrix} \dfrac{\partial t_1}{\partial x} & \dfrac{\partial t_1}{\partial y} & \dfrac{\partial t_1}{\partial z} & 1 \\ \dfrac{\partial t_2}{\partial x} & \dfrac{\partial t_2}{\partial y} & \dfrac{\partial t_2}{\partial z} & 1 \\ \vdots & \vdots & \vdots & \vdots \\ \dfrac{\partial t_n}{\partial x} & \dfrac{\partial t_n}{\partial y} & \dfrac{\partial t_n}{\partial y} & 1 \end{bmatrix}, \qquad B = \begin{bmatrix} t_{o1} - t_{c1} \\ t_{o2} - t_{c2} \\ \vdots \\ t_{on} - t_{cn} \end{bmatrix}.$$

(5)

Modification vector is got by Gaussian Elimination (4):

$$A^T A\Delta\theta = A^T B,$$

(6)

$$\Delta\theta = \left(A^T A\right)^{-1} A^T B.$$

(7)

After correction vector is obtained by (7), use $(\theta + \Delta\theta)$ as the new test point to keep iteration continuous until the error satisfies the requirements.

SMALL SCALE DIRECT SHEAR TESTS WITH DIFFERENT ROCK BRIDGE LENGTH

Many scholars have studied macrofailure and micropropagation of cracks evolution under shear loading, such as simulating microdevelopment of crack using cellular automata [24, 25], time distribution features of rock crack [26], brittle failure mechanism of rock [27], the influence of size effect on rock failure [28], and differential resistance sensing technology [29]. But small scale direct shear tests of different rock bridge length using AE are few.

As shown in Figure 3, the points on the failure plane are directly sheared to failure in sliding process. From a microscopic view point, the failure process of rock bridge of landslide is the direct shear failure. So rock bridge direct shear tests are carried out in order to study the failure mechanism with different rock bridge length in this paper.

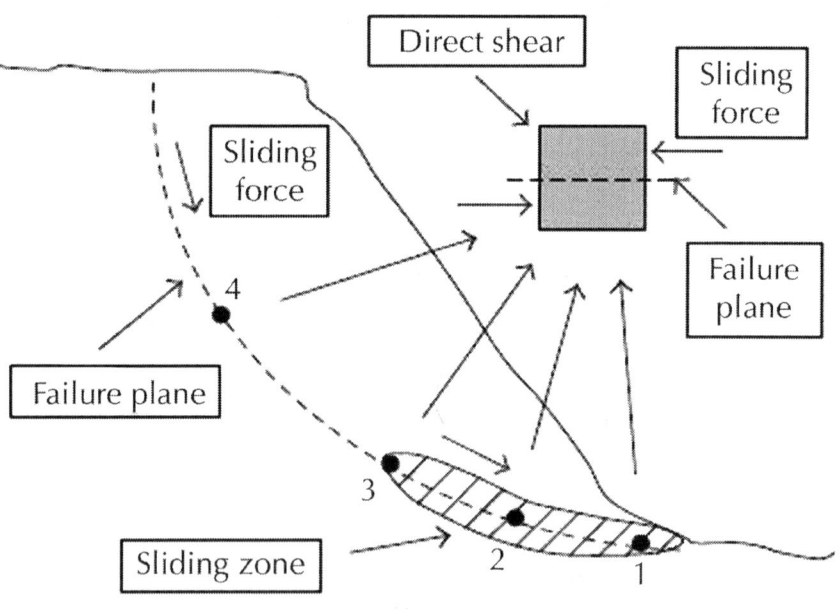

Figure 3: Sketch map of rock slope failure.

Sample Preparation and Test Instrument

The mass ratio of test sample is 1 : 3 : 1 : 1 (cement : sand : water : gypsum). According to the requirements of test instrument, the samples were processed into 50*50*50 mm³. The samples were divided into 4 groups and each sample is 25 mm, 30 mm, 35 mm, and 40 mm, respectively (Figure 4).

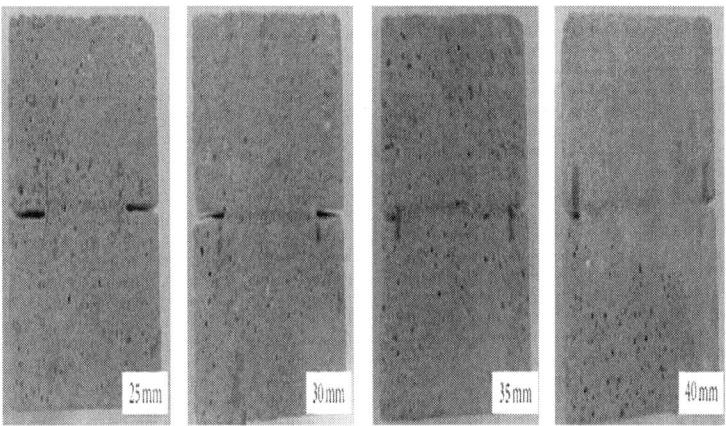

Figure 4: Different reserved rock bridge length.

The tests make use of two sets of devices that would load control system and AE monitoring system (Figure 5). The test instrument can be used to reasonably simulate rock stress condition, obtain the accurate parameters of shear strength, and complete stress-strain curve of sample. AE monitoring system is Micro-II Digital AE System which can be used to conduct real-time monitoring and record the whole process of the test. AE sensor arrangement is shown in Figure 6.

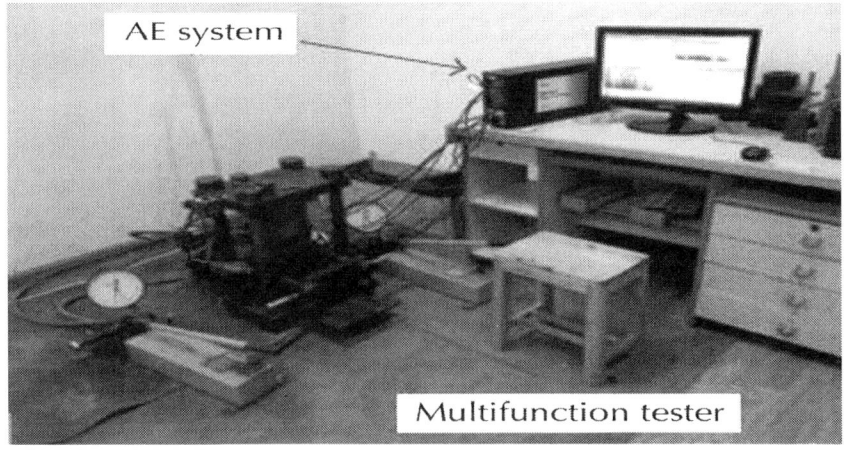

Figure 5: Direct shear test system associate with AE system

Figure 6: AE sensor arrangement.

Test Method

The tests were divided in four groups of 16 samples. The sample was loaded in sample box, while AE sensor was attached to the surface of sample box by means of coupling agent. Ready after all, the sample was applied positive pressure by means of normal jack and pressure values of each set of 4 samples under the vertical pressure were 0.4 MPa, 0.6 MPa, 0.8 MPa, and 1.0 MPa, respectively. Positive pressure was applied after completion and then shear stress was applied by using the horizontal jack. AE monitoring system came to work at the same time. Horizontal pressure was applied until the sample failure and pressure gauge data were recorded in order to obtain horizontal shearing forces during the test. AE monitoring system records the AE activity of rock automatically.

Test Results and Analysis

The Characteristics of AE Event Counts

To explore the AE change pattern under different rock bridge length and different vertical pressure, the AE event count with time graph is obtained under different rock bridge lengths of sample including

25 mm, 30 mm, 35 mm, and 40 mm under 0.6 MPa vertical pressures (Figure 7), and the AE event count is obtained when vertical pressure of sample is 0.4 MPa, 0.6 MPa, 0.8 MPa, and 1.0 MPa with 30 mm rock bridge length (Figure8). The AE event peak count with the rock bridge length and the vertical pressure graph are also mapped (Figures 9 and 10).

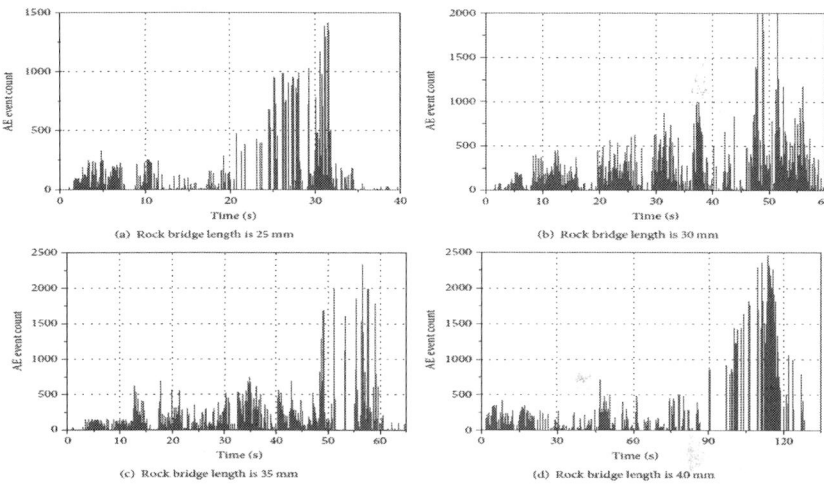

Figure 7: Relationship of AE event count and time at 0.6 MPa normal pressure under different rock bridge lengths.

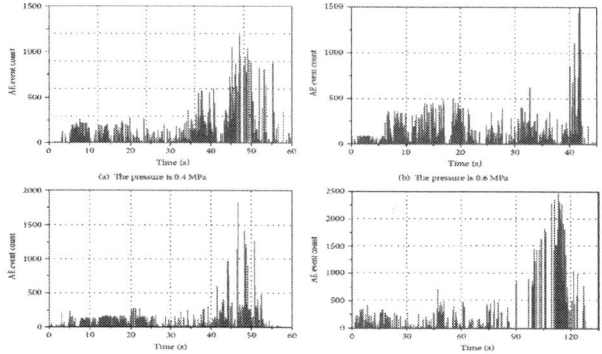

Figure 8: Relationship of AE event count and time at 30 mm rock bridge length under different vertical pressure.

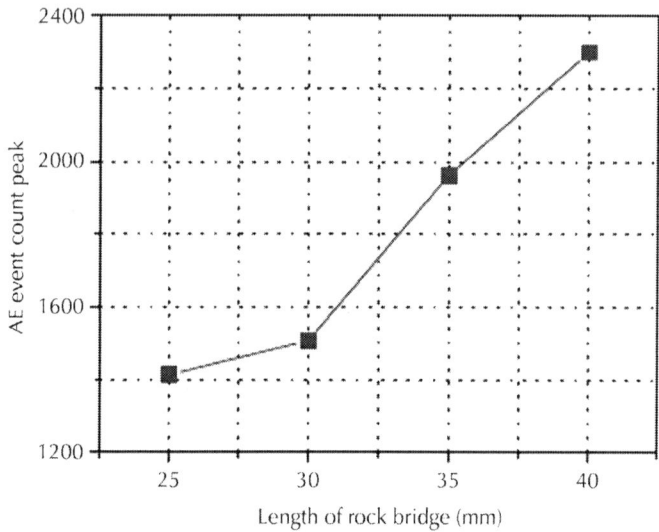

Figure 9: Relationship curves between AE event count peak and rock bridge length.

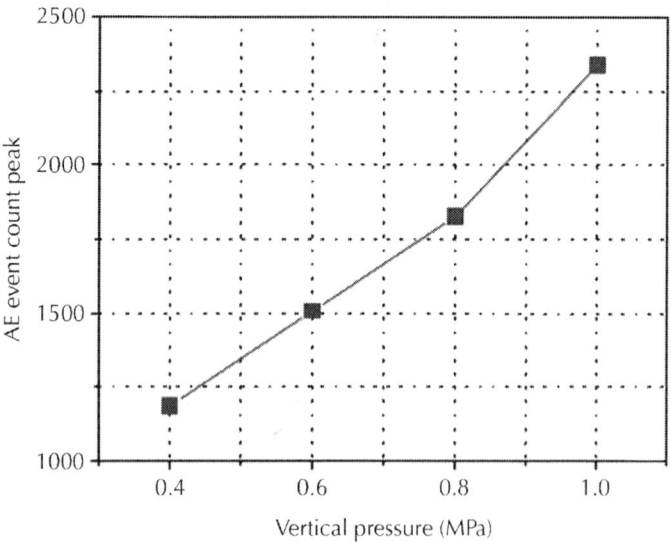

Figure 10: Relationship curves between AE event count peak and vertical pressure.

It can be seen that the sample failure mechanism has the following characteristics.

- In Figures 7 and 9, AE activity is more frequent and the appearance time of the AE event count peak value also increases with the increase of rock bridge length. This is because the arrival time of peak value increases with shear area under the same conditions, while shear area gains with rock bridge length.

- In Figures 8 and 10, AE event counts with the time have the same change trend; AE activity is more frequent. With the increasing of vertical pressure, AE event count peak value occurs much later. Shear stress increases with normal stress based on Mohr-Coulomb yield criterion, so the appearance time of peak value delays with the increase in vertical pressure.

Characteristics of AE Three-Dimensional Location

Figure 11 is the AE source location and the actual failure under different rock bridge lengths of 25 mm, 30 mm, 35 mm, and 40 mm under 0.6 MPa vertical pressure. Figure 12 is the AE source location of samples and the actual failure under different vertical pressure, consisting of 0.4 MPa, 0.6 MPa, 0.8 MPa, and 1.0 MPa under 30 mm length. Three-dimensional location result is shown by a side of sample in a picture to make AE three-dimensional location patterns more clear, and also it leads to some location points which overlap with each other. It can be found that the AE source location and the macroscopic failure mode of rock are in good match in Figures 11 and 12. The AE source location also has the following characteristics.

- In Figure 11, the number of AE location points was obtained with the increase of the rock bridge length. The crack shear angle increases from the horizontal direction to a certain angle, and the number of microcracks increases in the same vertical pressure under the different rock bridge length. It seems that location points with rock bridge of 30 mm break less than those with 25 mm in Figure 11.

- In Figure 12, the number of AE location points increases and its distribution is comparatively dispersive with the increase of vertical pressure; the crack shear angle was obtained from the horizontal direction to a certain angle in different vertical pressure under the same rock bridge length. The sample is easier to break with the increase of vertical pressure.

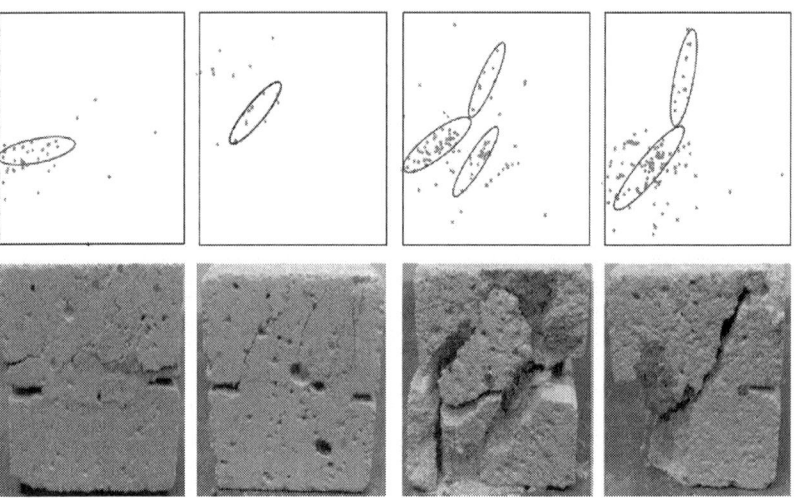

Figure 11: AE source locations of sample and the actual failure under different rock bridge length (25 mm, 30 mm, 35 mm, and 40 mm).

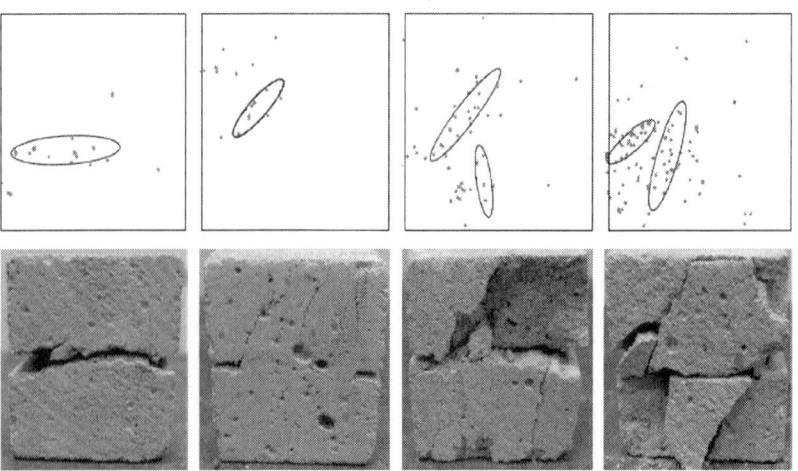

Figure 12: AE source locations of sample and the actual failure under different vertical pressure (0.4 MPa, 0.6 MPa, 0.8 MPa, and 1.0 MPa).

Failure Mechanism and Characteristics of Samples

As shown in Figure 13, it can be found that failure process can be divided into three stages by strain value.

- Stage 1 is the stage where strain value is 1~2% in which the shear stress is less than the shear strength of rock. The shear stress is much higher at the tip of rock bridge and the distribution of stress on the shear surface is uniform. At this stage, the AE activity is less, and it remains at a lower level and increases slowly.

- Stage 2 is the stage where strain value is 2~3%; the sample enters the stage of plastic deformation and begins to develop cracks with the increase of shear stress. Because of the stress concentration at the tip of rock bridge, the cracks often generate firstly from the tip of the rock bridge and finally grow into the main crack, and there form several groups of larger cracks parallel to the main crack near the rock bridge. A sudden integral brittle fracture occurs along the main rupture surface of the sample when the shear stress reaches the shear strength of rock. At this stage, AE activity frequency increases rapidly and AE count peak appears when the sample shear stress reaches the maximum value.

- Stage 3 is the stage with strain value after reaching 3%. As the sample is further damaged, the shear stress decreases rapidly and cracks extend further. At the end, the sample is completely cut out. At this stage, AE activity decays rapidly after maximum shear stress value and this stage is often shorter than others.

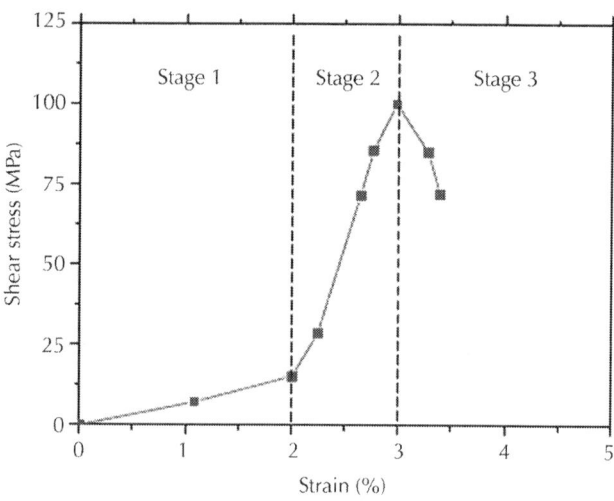

Figure 13: Sample stress-strain curve of direct shear test.

LARGE SCALE LANDSLIDE MODEL WITH LOCKED SECTION TEST

The locked section of rock slope refers to rock bridge near middle of sliding zone which has a relatively highly strength. The occurrence of large-scale rock landslides is generally accompanied by sudden brittle failure of the "locked section." It means that "locked section" plays an important role in deformation control and stability mechanism of rock slope, hence a key in assessment and control on slope disaster.

Some authors had investigated the micro- and macrocharacteristics of shear or tensile failure under the action of gravity creep by the view of creep mechanics of rock mass [30]. Some authors had researched the creep fracture degree of bedrock and distribution of serious landslide hazard areas by using seismic wave velocity test [31] and other methods [32]. The researches were carried out to study locked section in various aspects, but the AE technique is rarely used. In this paper, the dynamic test characteristics of AE are used to carry out the large scale landslide model test with locked section to have a further study on its failure mechanism.

Samples Preparation and Test Instruments

The mass ratio of test sample is set at 1:3:1:1 (cement:sand:water:gypsum), with the sample size shown in Figure 14(a). The locked section is circular arc and reserved length of 65 mm. The test system uses a multisystem integrated test method in order to achieve a better physical simulation condition. Loading platform is two-dimension geological system framework and it is 4 m long, upper beam 2.5 m high, and foot beam 0.3 m high. Hydraulic jack which contacts the model is fixed on the framework and controlled by hydraulic system. The jack provides different pressure to simulate the sliding of gravity according to the test requirement. The arrangement of AE sensors is shown in Figure 14(b). The bedrock boundary is fixed with the sliding mass boundary being set free. Vertical load is imposed on upper part of sliding mass.

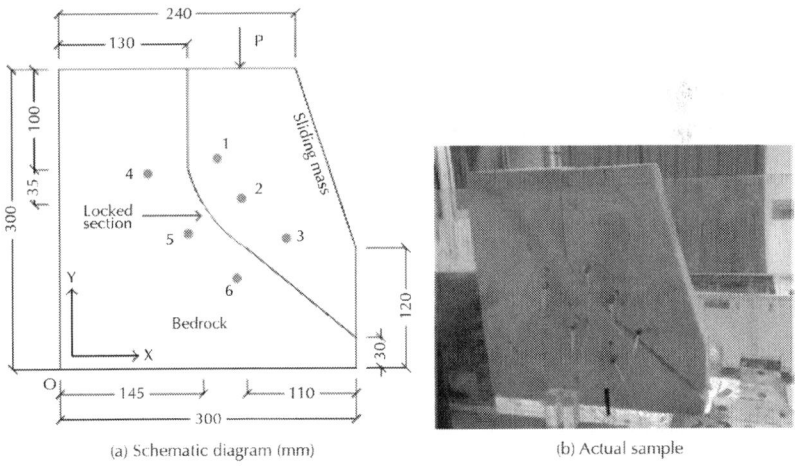

(a) Schematic diagram (mm)   (b) Actual sample

Figure 14: Sample size and sensor position.

Test Method

After the installing of AE system and sensors, then starting the hydraulic control system, controlling oil pump to create pressure, and taking gradual loading increase by 2 MPa every level, the gradual loading changes to 1 MPa every level when the crack propagation is obvious in locked section until the model failure during the end loading phase. The AE keep monitoring and recording during the whole test process.

Test Results and Analysis

The Characteristics of AE Event Count and AE Energy

According to test data, the AE event count and AE energy change with time during test progress of landslide model with the locked section are shown in Figure 15. The following can be concluded from Figure 15.

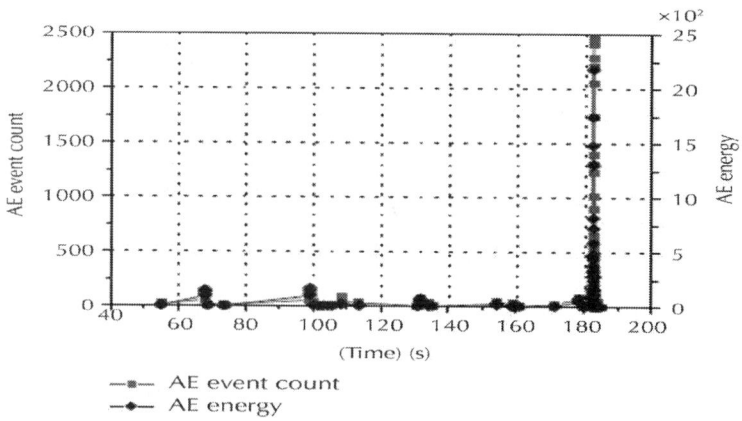

Figure 15: Relationship curves of AE event count with time and AE energy with time.

It can be seen from Figure 15 that the test is not like the direct shear test. The energy of locked section accumulates continuously because of the increase of stress. The energy suddenly releases and then rock mass of locked section is cut out. This process has the characteristic of sudden and brittle failure; it reflects the development process of rock slope that from the progressive damage evolutes to brittle failure. AE event count with time and AE energy with time of slope model with locked section are almost the same as each other where both have the same change trend and characteristics. It can be corroborated to well respond to the deformation and failure characteristics of the slope model.

Characteristics of AE Plane Location

The failure process and characteristics of locked section test model are examined through the comparison of Figure 16; the conclusion is as follows.

- In Figure 16, the failure surface is basically in the predetermined locked section; along with locked section, an inclined plane is formed with the right being high and the left being low. The length of shear failure section is about 40 mm, and the shear failure surface has obvious scratches. The tensile failure section is 20 mm, and tension cracks can be observed obviously (Figure 16(a)).

- It can be found in Figure 16(b) that the AE source location is in accordance with the actual failure. Analysis of AE source location process shows that the location point first appears in the outside of sensor array, and its position finally locates within the two-sensor array as AE source points appear continuously with the time. The location point is more intensive at the edge of locked section because of the stress concentration.

- From the perspective of overall effect of AE source and its location technology, AE source located accurately and the space-time evolution of cracks in locked section is reflected; the failure process and characteristics of slope during the localization process can also be demonstrated. Rock landslide with locked section has the characteristics such as higher energy accumulation, the sudden release of energy, and brittle failure, which cause long sliding distance and serious failure.

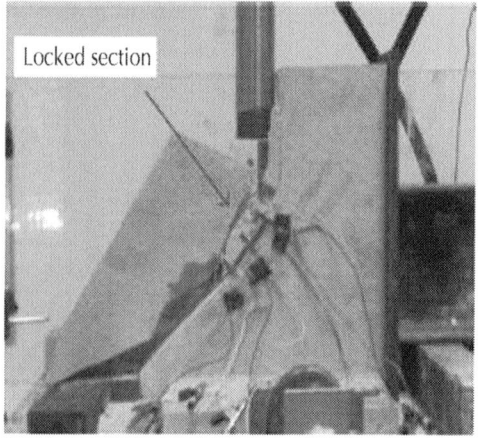 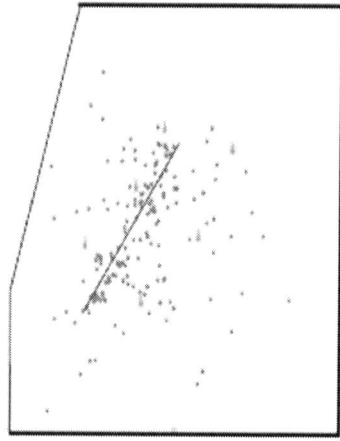

(a) Failure of locked section   (b) AE source location of failure

Figure 16: Failure of locked section under continuous loading.

CONCLUSIONS

In this paper, small scale direct shear tests with different lengths and large scale rock landslide model with locked section test are carried out; both of them aim to study the failure mechanism and mechanics characteristics of rock bridge of landslide. The AE event count and AE source location under different circumstances are analyzed. The conclusions can be summarized as follows.

- AE source location can be accurately located during small scale tests in small scale direct shear tests. the AE event count peak value increases with the increasing of rock bridge length and vertical stress. In addition, the time of AE event count peak value appearing goes back with the increase of the length and the vertical stress.

- AE resource location was accurately and reliably located during large scale landslide model test with locked section. Based on the locating process, rock sample failure features and cracks time-space evolution process were revealed. AE showed that the brittle failure feature of landslide with locked section was reflected well.

- AE technology plays an important role in rock mechanics and related research fields. The two tests based on AE technique well revealed the rock failure mechanism in rocky slope and clarified the reason of high speed and long distance sliding of rocky slope that cause serious disasters and catastrophe.

ACKNOWLEDGMENTS

This work is supported by the National Basic Research Program of China (973 Program, no. 2013CB733202) and the National Natural Science Foundation of China (Grant nos. 41272330 and 41130745). This work is also supported by the funding of Science and Technology Office of Sichuan Province (Grant no. 2015JQO020).

REFERENCES

1. D. Lockner, "The role of acoustic emission in the study of rock fracture," International Journal of Rock Mechanics and Mining Sciences and, vol. 30, no. 7, pp. 883–899, 1993.

2. N. W. Xu, C. A. Tang, L. C. Li et al., "Microseismic monitoring and stability analysis of the left bank slope in Jinping first stage hydropower station in Southwestern China," International Journal of Rock Mechanics and Mining Sciences, vol. 48, no. 6, pp. 950–963, 2011.

3. S. Li, X.-T. Feng, Z. Li, B. Chen, C. Zhang, and H. Zhou, "In situ monitoring of rockburst nucleation and evolution in the deeply buried tunnels of Jinping II hydropower station," Engineering Geology, vol. 137-138, pp. 85–96, 2012.

4. Q. Jiang, J. Cui, and J. Chen, "Time-dependent damage investigation of rock mass in an in situ experimental tunnel," Materials, vol. 5, no. 8, pp. 1389–1403, 2012.

5. V. Rudajev, J. Vilhelm, and T. Lokajíček, "Laboratory studies of acoustic emission prior to uniaxial compressive rock failure," International Journal of Rock Mechanics and Mining Sciences, vol. 37, no. 4, pp. 699–704, 2000.

6. D. P. Jansen, S. R. Carlson, R. P. Young, and D. A. Hutchins, "Ultrasonic imaging and acoustic emission monitoring of

thermally induced microcracks in Lac du Bonnet granite," Journal of Geophysical Research, vol. 98, no. 12, pp. 22231–22243, 1993.

7.   S. Wang, R. Huang, P. Ni, R. P. Gamage, and M. Zhang, "Fracture behavior of intact rock using acoustic emission: experimental observation and realistic modeling," Geotechnical Testing Journal, vol. 36, no. 6, 2013.

8.   D.-S. Cheon, Y.-B. Jung, E.-S. Park, W.-K. Song, and H.-I. Jang, "Evaluation of damage level for rock slopes using acoustic emission technique with waveguides," Engineering Geology, vol. 121, no. 1-2, pp. 75–88, 2011.

9.   C. Li, O. Stephansson, and T. Savilahti, "Behavior of rock joints and rock bridges in shear testing," inProceedings of the International Symposium on Rock Joints, pp. 259–266, 1990.

10.   B. Shen and O. Stephansson, "Cyclic loading characteristics of joints and rock bridges in a jointed rock specimen," in Proceedings of the International Symposium on Rock Joints, pp. 725–729, 1990.

11.   R. H. C. Wong, K. T. Chau, P. M. Tsoi, and C. A. Tang, "Pattern of coalescence of rock bridge between two joints under shear testing," in Proceedings of the 9th International Congress on Rock Mechanics, pp. 735–738, 1999.

12.   S.-Q. Yang, H.-W. Jing, and S.-Y. Wang, "Experimental investigation on the strength, deformability, failure behavior and acoustic emission locations of red sandstone under triaxial compression," Rock Mechanics and Rock Engineering, vol. 45, no. 4, pp. 583–606, 2012.

13.   A. Ghazvinian, V. Sarfarazi, S. A. Moosavi, et al., "Analysis of crack coalescence in rock bridges using neural network," in Proceedings of the European Rock Mechanics Symposium, pp. 255–258, 2010.

14.   F.-M. Zhang, B.-H. Wang, Z.-Y. Chen, X.-G. Wang, and Z.-X. Jia, "Rock bridge slice element method in slope stability analysis based on multi-scale geological structure mapping," Journal of Central South University of Technology, vol. 15, no. 2, pp. 131–137, 2008.

15.   J. Kemeny, "Time-dependent drift degradation due to the progressive failure of rock bridges along discontinuities,"

International Journal of Rock Mechanics and Mining Sciences, vol. 42, no. 1, pp. 35–46, 2005.

16. W. Zhu, S. Li, R. H. C. Wong, K. T. Chau, and J. Xu, "A study of fracture mechanism and shear strength of rock bridges through analytical and model-testing methods," Key Engineering Materials, vol. 261–263, pp. 225–230, 2004.

17. S. Y. Wang, W. Sloan, H. Y. Liu, and C. A. Tang, "Numerical simulation of the rock fragmentation process induced by two drill bits subjected to static and dynamic (impact) loading," Rock Mechanics and Rock Engineering, vol. 44, no. 3, pp. 317–332, 2011.

18. F. Dai, M. D. Wei, N. W. Xu, Y. Ma, and D. S. Yang, "Numerical assessment of the progressive rock fracture mechanism of cracked chevron notched Brazilian disc specimens," Rock Mechanics and Rock Engineering, 2014.

19. Z. Z. Liang, H. Xing, S. Y. Wang, D. J. Williams, and C. A. Tang, "A three-dimensional numerical investigation of the fracture of rock specimens containing a pre-existing surface flaw," Computers and Geotechnics, vol. 45, pp. 19–33, 2012.

20. H. G. Li, R. Zhang, M. Z. Gao, G. Wu, and Y. F. Zhang, "Advances in technology of acoustic emission of rock," Chinese Journal of Underground Space and Engineering, vol. 9, pp. 1794–1804, 2013.

21. Y.-L. Ding, Y. Deng, and A.-Q. Li, "Advances in researches on application of acoustic emission technique to health monitoring for bridge structures," Journal of Disaster Prevention and Mitigation Engineering, vol. 30, no. 3, pp. 341–351, 2010.

22. L. Geiger, "Probability method for the determination of earthquake epicenters from the arrival time only," Bulletin of St. Louis University, vol. 8, pp. 60–71, 1912.

23. W. Spence, "Relative epicenter determination using P-wave arrival-time differences," Bulletin of the Seismological Society of America, vol. 70, no. 1, pp. 171–183, 1980.

24. P.-Z. Pan, F. Yan, and X.-T. Feng, "Modeling the cracking process of rocks from continuity to discontinuity using a cellular automaton," Computers & Geosciences, vol. 42, pp. 87–99, 2012.

25. X. T. Feng, P. Z. Pan, and H. Zhou, "Simulation of the rock microfracturing process under uniaxial compression using an

elasto-plastic cellular automaton," International Journal of Rock Mechanics and Mining Sciences, vol. 43, no. 7, pp. 1091–1108, 2006.

26. X.-T. Feng and M. Seto, "Fractal structure of the time distribution of microfracturing in rocks,"Geophysical Journal International, vol. 136, no. 1, pp. 275–285, 1999.

27. G. Chen, T. Li, G. Zhang, H. Yin, and H. Zhang, "Temperature effect of rock burst for hard rock in deep-buried tunnel," Natural Hazards, vol. 72, no. 2, pp. 915–926, 2014.

28. D. Y. Li, C. C. Li, and X. B. Li, "Influence of sample height-to-width ratios on failure mode for rectangular prism samples of hard rock loaded in uniaxial compression," Rock Mechanics and Rock Engineering, vol. 44, no. 3, pp. 253–267, 2011.

29. G. Zhao, H. Pei, and H. Liang, "Measurement of additional strains in shaft lining using differential resistance sensing technology," International Journal of Distributed Sensor Networks, vol. 2013, Article ID 153834, 6 pages, 2013.

30. M. Chigira, "Long-term gravitational deformation of rocks by mass rock creep," Engineering Geology, vol. 32, no. 3, pp. 157–184, 1992.

31. N. W. Xu, F. Dai, Z. Z. Liang, Z. Zhou, C. Sha, and C. A. Tang, "The dynamic evaluation of rock slope stability considering the effects of microseismic damage," Rock Mechanics and Rock Engineering, vol. 47, no. 2, pp. 621–642, 2014.

32. G. S. Zhao, G. Q. Zhou, X. D. Zhao, Y. Z. Wei, and L. J. Li, "R/S analysis for stress evolution in shaft lining and fracture prediction method," Advanced Materials Research, vol. 374–377, pp. 2271–2274, 2012.

Chapter 9

Coupling Local and Non-local Damage Evolutions with the Thick Level Set Model

Nicolas Moës[1], Claude Stolz[1, 2],
and Nicolas Chevaugeon[1]

[1]Ecole Centrale de Nantes, GeM Institute, UMR CNRS 6183, 1 Rue de la Noe, Nantes 44321, France

[2]Lamsid, EDF-CEA-CNRS UMR 2832, Avenue du Général de Gaulle, Clamart 92141, France

ABSTRACT

Background

The Thick Level Set model (TLS) is a recent method to delocalize local constitutive models suffering spurious localization. It has two major advantages compared to other delocalization methods. The first one is that the transition from localization to fracture is taken into account in the model. The second one is that the delocalization only acts when and where needed. In other words, the TLS has no effect when the local model is stable. The former advantage was already detailed in several papers (IJNME 86:358-380, 2011, CMAME 233:11-27, 2012, IJF 174:49-60, 2012). This paper concentrates on the latter advantage.

Methods

The TLS delocalization approach is formulated as a bound on the damage gradient. The non-local zone is defined as the zone where the bound is met whereas the local zone is defined as the zone where it is not met. The boundary (localization front) between the local and non-local zone is the main unknown in the problem.

Results

Based on the new model, a 1D pull-out test is solved both analytically and numerically. Different regimes are observed in the solution as the loading progresses: fully elastic, local damage, coupled local/non-local damage and, finally, purely non-local damage.

Conclusions

The new model introduces delocalization as an inequality allowing local damage to develop in zones whereas non-local damage may develop in other zones. This reduces dramatically the cost of implementation of such models compared to fully non-local models.

BACKGROUND

Although the scope of TLS application is much wider, we consider in this paper the fracture of quasi-brittle structures under quasi-static loading and under small deformation assumption. The loading is proportional to a scalar parameter. The material is modelled by a time-independent elasto-damage constitutive model with scalar damage.

Due to quasi-static analysis, the loading parameter must be controlled especially when bifurcation occurs.

The TLS model was introduced in several papers [1],[2] and [3],[4]. It lies between continuum damage mechanics and fracture mechanics. Indeed, crack opening is allowed across fully damaged zones (see [2] for instance). The fully damaged zone is located by a level set. Let us note that the description above is different from a diffuse vision of the crack in which crack opening is not explicitly modeled as in the phase-field approach [5]-[7] or the variational approach to fracture [8],[9]. We are rather in the vein of transition from damage to fracture as in [10]. However, the TLS will not be in need of a cohesive zone to perform the transition. The model can be considered as a continuous transition from damage to fracture.

The main idea of the TLS for quasi-brittle fracture is to bound the spatial gradient of the damage variable d, thus avoiding spurious localization. One imposes that the spatial damage distribution satisfies at all time

$$\|\nabla d\| \leq f(d) \text{ on } \Omega$$

(1)

Where Ω is the domain of interest. The choice of the function $f(d)$ will be discussed in what follows. As damage evolves, one eventually wants to locate the crack, i.e. the zone for which $d=1$. However, finding the iso-contour $d=1$ for a quantity d than cannot go beyond 1 is a tedious operation. This is where the level set ingredient comes into play. Variable d is expressed in terms of a level set as depicted in Figure 1. This relation introduces a length scale l_c. Finding the zone $d=1$, is now well-posed since the level set is not strictly limited to l_c but may go beyond. With the use of the surrogate variable , condition (1) may be rewritten as

$$\begin{cases} \|\nabla\phi\| \leq 1 \\ d = d(\phi) \end{cases}$$

(2)

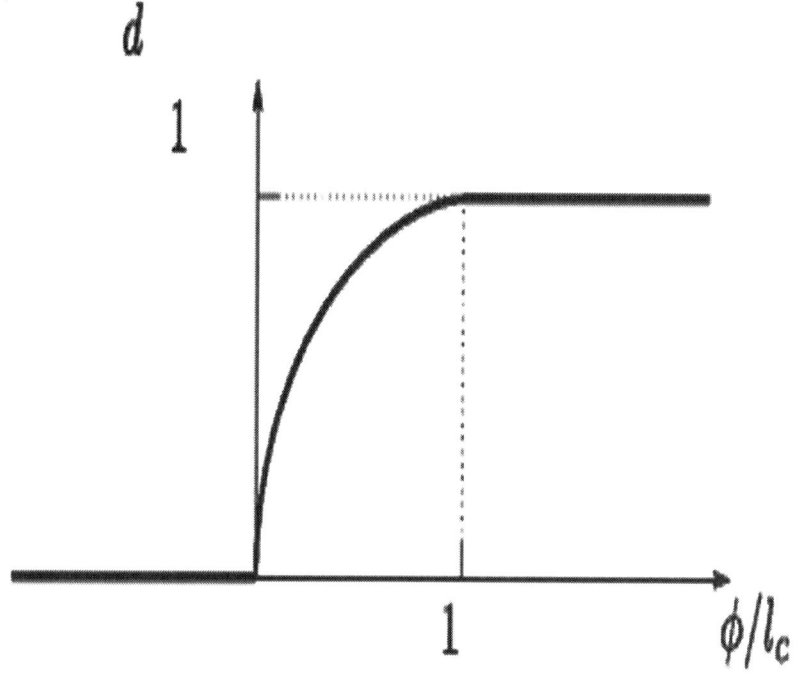

Figure 1: An example of damage shape function.

where $f(d)$ in (1) is related to $d((\phi))$ by $f(d)=d'(\phi\ (d))$ (the prime indicating the derivative of d with respect to ϕ). The function $d\ (\phi)$ is called the damage shape function and is the main ingredient of the TLS. Equation (2) above indicates that ϕ is a distance function in the zone where the constraint is active (we name this zone the localization zone). The evolution of a distance function has been analyzed and updating algorithm proposed in [11]. In the localization zone, the evolution of ϕ is non-local, indeed

$$\|\nabla\phi\| = 1 \Rightarrow \nabla\dot{\phi}\cdot\nabla\phi = 0$$

(3)

The rate of change of ϕ is thus uniform on any segment aligned with $\nabla\phi$ and the rate of d is given by $d^{\cdot} = d{\cdot}(\phi)\dot{\phi}$. Such segments over which $\dot{\phi}$ is uniform are depicted in Figure 2. In the local zone, the evolution of ϕ stems from the evolution of d and the relation $d = d(\phi)$.

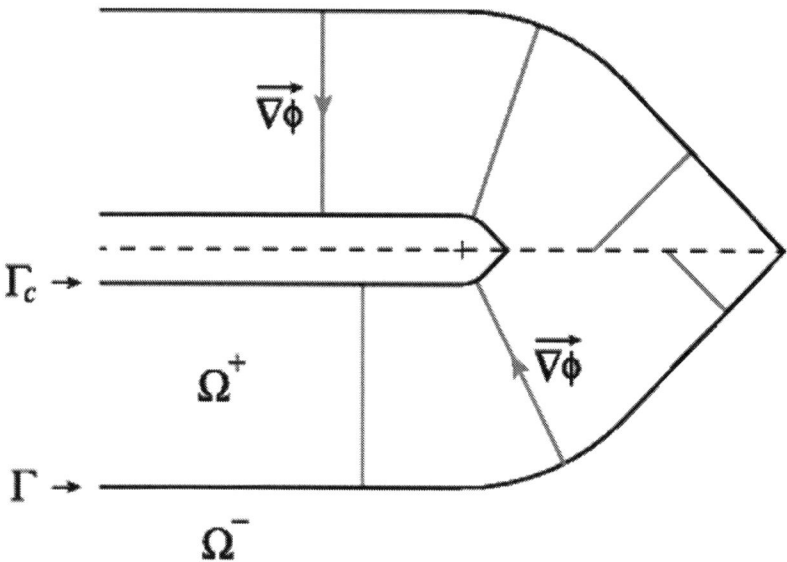

Figure 2: Local (Ω^-) and localization (Ω^+) domains as well as cracks faces (G_c). Inside Ω^+, we have $\|\nabla\phi\| = 1$. It can be noted that $\nabla\phi$ (and thus ∇d) is discontinuous along the dashed line which is the so-called skeleton of the distance function. The

The delocalization (1) used in the TLS is different from existing delocalization techniques. Indeed, it directly uses the norm of the damage gradient. It is thus a Hamilton-Jacobi type equation. On the contrary, damage gradient models [12]-[14] yield Laplacian damage type equation rising the question of proper boundary conditions.

The TLS shares some similarities with the so-called non-local integral approach [15],[16] in which weighted averages are performed over segments (1D), disks (2D) and spheres (3D) of fixed size. In the TLS approach, however, weighted averages are always performed on segments (Figure 2) whatever the dimension of the body and over a length which is not fixed in time but evolves from zero to a maximum length l_c. Finally, note that as l_c is the minimal distance between a point where $d=0$ and a fully damaged point, $d=1$, it plays the role of the fracture process zone size.

After this quick introduction of the TLS, we get to the objective of the paper. In previous TLS paper, the delocalization condition (2) was considered as an equality on the whole domain. It meant that d was zero on the domain except in zone where the gradient norm was fixed. The short-coming of this view was that uniform or smooth damage field (because of damage hardening for instance) could not be modeled prior to localization. The inequality analyzed in this paper allows a combination of local and non-local evolutions. In the literature, the possibility to combine both local and non-local approach is seldom discussed with the exception of the so-called morphing numerical technique [17],[18].

The paper is organized as follows. The TLS concept with the inequality constraint discussed above are detailed in the first section. Next, the TLS boundary value problem is set up and a dissipation analysis is carried out. A 1D pull-out is solved semi-analytically to show the main feature of the TLS solution. This 1D test is then solved numerically with the TLS to observe the influence of the parameters choice in the model. A conclusion and perspectives end the paper.

METHODS

We consider a solid body occupying a domain Ω. The external surface $\partial \Omega$ is composed of two parts ∂W_u and ∂W_T on which the displacements lu^d and the loading lT^d are prescribed, respectively. The parameter λ is a loading parameter.

Small strains and displacements are assumed as well as quasi-static evolution. The current state is characterized by the displacement field

u, from which the strain field $\epsilon = \frac{1}{2}\left(\nabla u + \nabla u^T\right)$ is derived. The current state is also characterized by an internal scalar variable, the damage denoted d. In this paper, we will not consider other internal variables. Regarding the material model, we consider a free energy $\psi(\epsilon, d)$ from which the stress tensor σ and local energy release rate Y may be derived

$$\sigma = \frac{\partial \psi(\epsilon, d)}{\partial \epsilon}, \quad Y = -\frac{\partial \psi(\epsilon, d)}{\partial d} \tag{4}$$

The potential ψ is assumed for now at least convex with respect to ϵ. The need for other properties will be discussed later. Regarding the time-independent damage evolution, we consider a function y depending on damage and strain history (through e) such that

$$\dot{d} \geq 0, \quad y(e,d) - Y_c \leq 0, \quad (y(e,d) - Y_c)\dot{d} = 0$$

(5)

where

$$e = e(\epsilon(\tau), \tau \leq t)$$

(6)

and Y_c is some threshold. We believe the above formalism encompasses most of the damage models in the literature. To be even more general, one may consider two relations of the kind (5): one for damage in tension and a second one for damage in compression and then combine these damages into d. One has a so-called associated damage model when the y variable is Y. In this case, damage evolution is expressed in terms of the dissipation potential $\varphi^*(Y)$ which is the indicator function of $Y - Y_c \leq 0$:

$$\dot{d} \in \frac{\partial \varphi^*(Y)}{\partial Y}$$

(7)

Such model was already considered in [1] for dissymmetric tension-compression evolution. We emphasize the fact that the TLS description is not restricted to associated damage models.

What we have described so far is a purely local damage model. This type of model is known to suffer spurious localizations meaning that the damage gradient may become infinite. The main idea of the TLS approach is to bound damage gradient as expressed in (1). In the TLS model, damage is allowed to go to 1 (but not beyond of course). The location of a crack (or fully degraded zones like in comminution problems) is defined by the set of points for which $d=1$. Numerically speaking, finding the set of points for which $d=1$ knowing that d may not go beyond 1 is not very practical. This is why the TLS expresses damage in terms of a surrogate variable ϕ whose values are not limited as depicted in Figure 1. We assume the following regularity on $d(\phi)$

$$\begin{cases} d(\phi) \in C^0(]-\infty, +\infty[) \text{ and monotonically increasing} \\ d(\phi) = 0 \text{ if } \phi \leq 0 \\ d(\phi) = 1 \text{ if } \phi \geq l_c \\ d(\phi) \in C^1(]0, l_c[) \end{cases}$$

(8)

Finding the subdomain where $d=1$ is equivalent to find the subdomain whose boundary is the iso-contour $\phi = l_c$.

In terms of the surrogate variable, ϕ, condition (1) reads

$$\|\nabla\phi\| \leq 1 \tag{9}$$

provided $f(d)$ is given by

$$f(d) = d'(\phi(d)) \tag{10}$$

For instance, if d is linear with respect to ϕ, the gradient of damage will be bounded by a constant

$$d = \phi/l_c, \phi \in [0, l_c] \implies \|\nabla d\| \leq \frac{1}{l_c} \tag{11}$$

Where as for more complex function $d(\phi)$, the bound depends on the level of damage. For instance, for the profile shown in Figure 1, we have

$$d = 1 - (1 - (\phi/l_c))^2 \implies \|\nabla d\| \leq \frac{2}{l_c}\sqrt{1-d} \tag{12}$$

For a general power law with $n \geq 1$, we obtain

$$d = 1 - (1 - (\phi/l_c))^n \implies \|\nabla d\| \leq \frac{n}{l_c}(1-d)^{1-1/n} \tag{13}$$

Whether local or non-local constitutive model should be used at a point x is based on condition (9).

$$\|\nabla\phi(x)\| < 1 \Rightarrow \text{Local constitutive model at } x \tag{14}$$

$$\|\nabla\phi(x)\| = 1 \Rightarrow \text{Non-Local constitutive model at } x \tag{15}$$

$$\|\nabla\phi(x)\| > 1 \qquad \text{forbidden} \tag{16}$$

The first condition is the major novelty of this paper, compared to previous paper on the TLS. At any time t, the domain may thus be decomposed into three non-overlapping zones : a local zone Ω^-, a non-local zone Ω^+ and a fully damaged zone Ω_c

$$\overline{\Omega} = \overline{\Omega_c} \cup \overline{\Omega^+} \cup \overline{\Omega^-} \tag{17}$$

$$\Omega^- = \{x \in \Omega : \|\nabla\phi(x)\| < 1, \phi(x) < l_c\} \tag{18}$$

$$\Omega^+ = \{x \in \Omega : \|\nabla\phi(x)\| = 1, \phi(x) < l_c\} \tag{19}$$

$$\Omega_c = \{x \in \Omega : \phi(x) \geq l_c\} \tag{20}$$

We define also the boundary G_c of the fully damaged zone and the interface Γ between the local and non-local zones.

$$\Gamma_c = \partial\Omega_c, \quad \Gamma = \overline{\Omega^+} \cap \overline{\Omega^-} = \partial\Omega^+ \cap \partial\Omega^- \tag{21}$$

The boundary G_c defines the crack faces. Figure 2 shows a typical scenario of a crack appearing inside the localization zone.

Note that the volume measure of Ω_c may be zero. This information is part of the solution process. We expect different shapes of W_c in comminution and brittle crack propagation.

Eikonal Equation

Condition, $\|\nabla\phi(x)\| = 1$ is a non-linear first-order partial differential equation. It is called an eikonal equation and belongs to the Hamilton-Jacobi equation family. Among the possible solution satisfying $\|\nabla\phi(x)\| = 1$, we will pick the one corresponding to the vanishing viscosity solution [19]. It is characterized by

$$\phi(x) = \min_{y \in \Gamma}(\phi(y) + d(x, y)), x \in \Omega^+ \tag{22}$$

where $d(x,y)$ is the length of the shortest path connecting x and y inside Ω^+. The value of ϕ at $x \in \Omega^+$ can be thought as the minimal fare to go from Γ to x. The fare being the sum of the initial fare $\phi(y)$ plus the mileage from y to x. Damage on Ω^+ is thus fully determined from values on Γ. A 1D example of ϕ satisfying the eikonal on a segment [b,d] is given in Figure 3.

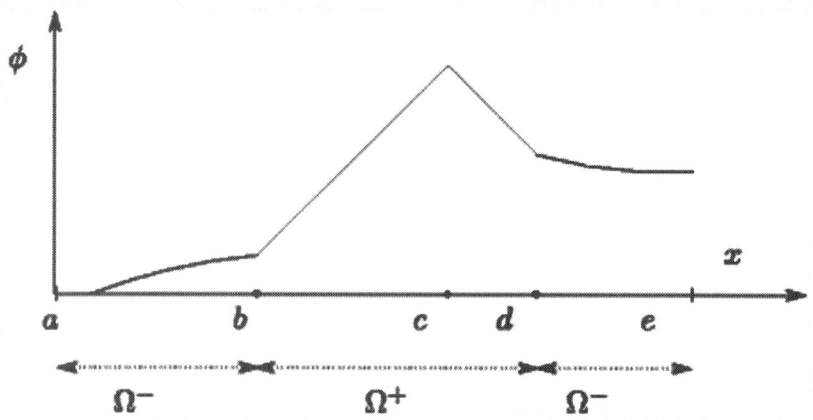

Figure 3: Distribution of ϕ on a 1D domain: $\Omega = [a,e]$, $\Omega^+ = [b,d]$, $\Omega^- = [a,b] \cup [d,e]$, $\Gamma = \{b,d\}$. Slopes at 45 degrees on Ω^+ indicate that ϕ behaves as a distance function ($\|\nabla\phi(x)\| = 1$) whereas $\|\nabla\phi(x)\| < 1$ on Ω^-. Point c is the skeleton of the distance function.

The fact that damage is related to a variable satisfying the eikonal equation, the cornerstone of the level set technology [11], explains why the damage model is coined Thick Level Set. In the non-local zone, damage is modeled over a thick layer in terms of level sets.

Damage Evolution

In the local zone, Ω^-, damage evolution is local and given by (5). In the non-local zone, Ω^+, damage rate is related to $\dot{\phi}$ by

$$\dot{d} = d'(\phi)\dot{\phi}$$

$$(23)$$

where $\dot{\phi}$ is uniform on segments aligned with $\nabla\phi$, see Equation (3). We denote this space as A:

$$\dot{\phi} \in \mathcal{A} = \left\{ a(x) \in L^2(\Omega^+) : \nabla a \cdot \nabla\phi = 0 \right\}$$

$$(24)$$

Non-local damage evolution boils down to decomposing Ω^+ into a set of independent segments and finding a value $\dot{\phi}$ over each of them.

As in [2], we suggest to introduce averaged quantities, $\bar{y}, \bar{d}$ over each segments. This may be expressed by a projection operation.

$$\bar{y} \in \mathcal{A} : \int_{\Omega^+} \bar{y} d'a \; d\omega = \int_{\Omega^+} y d'a \; d\omega, \quad \forall a \in \mathcal{A} \tag{25}$$

$$\bar{\dot{d}} \in \mathcal{A} : \int_{\Omega^+} \bar{\dot{d}} a \; d\omega = \int_{\Omega^+} \dot{d} a \; d\omega, \quad \forall a \in \mathcal{A} \tag{26}$$

We note that the averages satisfy the following property

$$\int_{\Omega^+} \bar{y} \bar{\dot{d}} \; d\omega = \int_{\Omega^+} y \dot{d} \; d\omega \tag{27}$$

The above indicates that duality is preserved through the averaging technique. This is not often the case in delocalization techniques as discussed in [20].

The local constitutive model, (5), is then expressed in terms of the non-local quantities

$$\bar{\dot{d}} \geq 0, \quad \bar{y} - Y_c \leq 0, \quad (\bar{y} - Y_c)\bar{\dot{d}} = 0 \tag{28}$$

where we have assumed Y_c uniform (if not it needs to be averaged by formula (25)). Finally, we write the relation giving $\dot{\phi}$ in terms of $\bar{d}$:

$$\bar{\dot{d}} = \bar{d'} \dot{\phi}, \quad \bar{d'} \in \mathcal{A} : \int_{\Omega^+} \bar{d'} a \; d\omega = \int_{\Omega^+} d'a \; d\omega, \quad \forall a \in \mathcal{A} \tag{29}$$

To end this section we illustrate the average formula on the 1D example depicted in Figure 3. Averages are given by

$$\text{On } [b,c]: \bar{y} = \frac{\int_b^c y d'(\phi) \; dx}{\int_b^c d'(\phi) \; dx}, \quad \bar{\dot{d}}(x) = \frac{\int_b^c \dot{d} \; dx}{\int_b^c \; dx} \tag{30}$$

$$\text{On } [c,d]: \bar{y} = \frac{\int_c^d yd'(\phi)\,dx}{\int_c^d d'(\phi)\,dx}, \quad \bar{d}(x) = \frac{\int_c^d \dot{d}\,dx}{\int_c^d dx}$$

$$(31)$$

TLS Boundary Value Problem

We are now able to define the boundary value problem. The set of admissible displacements is given by

$$\mathcal{U} = \left\{ u \in C^0(\Omega \setminus \Omega_c), \int_{\Omega \setminus \Omega_c} \psi(\epsilon(u), d(\phi))\,d\omega < +\infty, u = \lambda u^d \text{ on } \partial\Omega_u \right\}$$

$$(32)$$

The fact that the fully damage zones are removed from the domain is important. It allows the displacement to be discontinuous across G_c. Regarding the regularity of the displacement, we request that the energy, i.e. integral of ψ over Ω/Ω_c, is finite. This space is not simply H^1 as in elasticity since the stiffness is possibly vanishing on G_c boundary [21].

Regarding the ϕ variable, it is required to be continuous over Ω and belong to the set κ. The admissible set for ϕ is denoted K.

$$K = \left\{ \phi \in C^0(\Omega) : \|\nabla\phi(x)\| = 1, x \in (\Omega^+ \cup \Omega_c), \|\nabla\phi(x)\| < 1, x \in \Omega^- \right\}$$

$$(33)$$

The continuity requirement on ϕ leads to a Hadamard compatibility condition on the moving boundary Γ. Let us define the jump of a quantity f across Γ by

$$[f]_\Gamma(x, t) = f^+(x, t) - f^-(x, t)$$

$$(34)$$

The exponent -/+ placed on some quantities f defined at x on Γ has the following meaning

$$f^\pm(x, t) = \lim_{h \to 0^+} f(x \pm hn(x, t), t)$$

$$(35)$$

where n is the outward normal to Ω^+. With these notations we have

$$[\dot{\phi}]_\Gamma + [\nabla\phi]_\Gamma \cdot \boldsymbol{n}\, v_n = 0$$

(36)

where v_n is the normal velocity of Γ counted positively along $\boldsymbol{n}$. This gives the respective evolution of domains Ω^+ and Ω^-.

Potential energy of the domain is given by :

$$\begin{cases} \boldsymbol{u} \in \mathcal{U}, \ \phi \in K \\ E^{\text{pot}}(\boldsymbol{u}, \phi, \lambda) = \int_{\Omega^-} \psi(\epsilon(\boldsymbol{u}), d(\phi))\, d\omega + \int_{\Omega^+} \psi(\epsilon(\boldsymbol{u}), d(\phi))\, d\omega \\ \qquad\qquad - \int_{\partial\Omega_T} \lambda T^{\text{d}} \cdot \boldsymbol{u}\, da \end{cases}$$

(37)

Note that the same free energy expression, ψ, is used over Ω^+ and Ω^-. In what follows, $\boldsymbol{n}$ is the outward normal vector to Ω on $\partial\Omega$ and to Ω^+ on $\partial\Omega^+$. The set of admissible displacements variations is denoted as u_0. It has the same definition as except that $\boldsymbol{u}$ is set to zero on ∂W_u.

Assuming, that at time t, the spatial distribution of ϕ of the two volumes Ω^+ and Ω^- is known, the displacement field $\boldsymbol{u}$ is the field that solves the stationarity of the potential energy:

$$E^{\text{pot}}_{,u}\,\delta u = 0, \qquad \forall \delta u \in \mathcal{U}_0$$

(38)

This means;

$$\int_{\Omega^-} \sigma : \epsilon(\delta u)\, d\omega + \int_{\Omega^+} \sigma : \epsilon(\delta u)\, d\omega - \int_{\partial\Omega_T} \lambda T^{\text{d}} \cdot \delta u\, da = 0, \quad \forall \delta u \in \mathcal{U}_0$$

(39)

For simplicity, we assume that the boundary G_c is traction free (no contact on crack faces). The equilibrium (39) yields the following local equations

$$\text{div}\,\sigma = \boldsymbol{0} \ \text{ over } \Omega \setminus \Omega_c$$

(40)

$$[\sigma]_\Gamma \cdot \boldsymbol{n} = \boldsymbol{0} \ \text{ on } \Gamma$$

(41)

$$\sigma \cdot \boldsymbol{n} = \lambda T^{\text{d}} \ \text{ on } \partial\Omega_T$$

(42)

$$\sigma \cdot \boldsymbol{n} = \boldsymbol{0} \ \text{ on } \Gamma_c$$

(43)

We stress the fact that G_c denotes the boundary of the fully damage zone and thus in case of a crack G_c indicates *both* crack lips. To complete the set of equations to be solved for a known damage distribution, we add the stress definition and kinematic relations

$$\sigma = \frac{\partial \psi(\epsilon, d)}{\partial \epsilon} \quad \text{over } \Omega \setminus \Omega_c$$

(44)

$$\epsilon = \frac{1}{2}\left(\nabla u + \nabla u^T\right) \quad \text{over } \Omega \setminus \Omega_c$$

(45)

$$[u]_\Gamma = 0 \quad \text{on } \Gamma$$

(46)

$$u = \lambda u^d \quad \text{on } \partial \Omega_u$$

(47)

Finally, we need to add damage evolution equations in the local zone (5) and non-local zone (28).

Dissipation Analysis and Fields Regularity

The goal of this section is to analyze the expression of the dissipation as well as looking at the fields regularity across the boundary Γ.

Taking into account the conservation law for the total energy during the evolution of the system, the total dissipation associated with the loading rate $\dot{\lambda}$ is:

$$D = \int_{\partial \Omega_T} \lambda T^d \cdot \dot{u} \, da + \int_{\partial \Omega_u} \dot{\lambda} u^d \cdot \sigma \cdot n \, da$$

(48)

$$-\frac{d}{dt}\left(\int_{\Omega^-} \psi(\epsilon, d(\phi)) \, d\omega + \int_{\Omega^+} \psi(\epsilon, d(\phi)) \, d\omega\right)$$

(49)

In the above, we did not consider the energy inside W_c because it is assumed to be zero. Indeed, no compression is considered in this zone (see (43)).

Using Leibniz formula for the time derivative of moving domains as well as the relation:

$$\dot{\psi} = \sigma : \epsilon(\dot{u}) - Y\dot{d}$$

(50)

we obtain

$$D = \int_{\partial\Omega_T} \lambda T^d \cdot \dot{u}\, da + \int_{\partial\Omega_T} \lambda u^d \cdot \sigma \cdot n\, da - \int_{\Omega^-} \sigma : \epsilon(\dot{u})\, d\omega - \int_{\Omega^+} \sigma : \epsilon(\dot{u})\, d\omega$$
$$+ \int_{\Gamma} [\psi]_{\Gamma}\, v_n\, da - \int_{\Gamma_c} \psi\, v_n\, da + \int_{\Omega^-} Y\dot{d}\, d\omega + \int_{\Omega^+} Y\dot{d}\, d\omega$$

Integrating the domain integral by parts in the second line above and using the equations characterizing the equilibrium state, we get

$$D = \int_{\Gamma} [\psi]_{\Gamma}\, v_n + n \cdot \sigma \cdot [\dot{u}]_{\Gamma}\, da - \int_{\Gamma_c} \psi\, v_n\, da + \int_{\Omega^-} Y\dot{d}\, d\omega + \int_{\Omega^+} Y\dot{d}\, d\omega$$

(51)

During the propagation of the interface, perfect contact is assumed on Γ, that is the displacement jump across Γ must be zero at all time. As a consequence, the derivative along the moving interface of the displacement jump must be zero, [22], yielding the so called first Hadamard compatibility condition between the front velocity v_n and the jump in material velocities $[\dot{u}]_{\Gamma}$:

$$[\dot{u}]_{\Gamma} + [\nabla u]_{\Gamma} \cdot n\, v_n = 0$$

(52)

Equation (51), now becomes

$$D = \int_{\Gamma} n \cdot [P]_{\Gamma} \cdot n\, v_n\, da - \int_{\Gamma_c} \psi\, v_n\, da + \int_{\Omega^-} Y\dot{d}\, d\omega + \int_{\Omega^+} Y\dot{d}\, d\omega$$

(53)

where P is the Eshelby tensor

$$P = \psi I - \sigma \cdot \nabla u$$

(54)

The first term is the dissipation created by the interface propagation. We show now that due to damage continuity on Γ this term is zero.

Since normal stress and displacement are continuous across Γ, the product of the jump in stress and strain across Γ is zero, [23],[24]:

$$[\sigma]_{\Gamma} : [\epsilon]_{\Gamma} = 0$$

(55)

Let $y_d(\epsilon)$ be the density of free energy for a given value of damage and let $\psi^*_d(\sigma)$ be its dual by the Legendre-Fenchel transform. Since the couples (ϵ^+,σ^+) and (ϵ^-,σ^-) do satisfy the constitutive model (4), we have

$$\psi_d(\epsilon^+) + \psi_d^*(\sigma^+) - \sigma^+ : \epsilon^+ = 0, \quad \psi_d(\epsilon^-) + \psi_d^*(\sigma^-) - \sigma^- : \epsilon^- = 0$$

(56)

Summing the two relations above and using (55), we have

$$\left(\psi_d(\epsilon^+) + \psi_d^*(\sigma^-) - \sigma^- : \epsilon^+\right) + \left(\psi_d(\epsilon^-) + \psi_d^*(\sigma^+) - \sigma^+ : \epsilon^-\right) = 0$$

(57)

Since both terms above are greater or equal to zero (classical property of convex analysis, see [25]), we have

$$\psi_d(\epsilon^+) + \psi_d^*(\sigma^-) - \sigma^- : \epsilon^+ = 0, \quad \psi_d(\epsilon^-) + \psi_d^*(\sigma^+) - \sigma^+ : \epsilon^- = 0$$

(58)

This implies that the couples (ϵ^+,σ^-) and (ϵ^-,σ^+) do satisfy the constitutive model. Assuming that the convex potential $y_d(\epsilon)$ is such that the stress associated to any strain is unique, we have

$$[\sigma]_\Gamma = [\epsilon]_\Gamma = 0$$

(59)

The continuity of the strain and displacement across Γ leads to the continuity of the displacement gradient

$$[\nabla u]_\Gamma = 0$$

(60)

leading finally to the continuity of the Eshelby tensor.

$$[P]_\Gamma = 0$$

(61)

Dissipation is thus reduced to

$$D = \int_{\Omega^-} Y \dot{d} \, d\omega + \int_{\Omega^+} Y \dot{d} \, d\omega - \int_{\Gamma_c} \psi \, v_n \, da$$

(62)

The dissipation must be positive. For classical models in which Y is positive, this implies that damage may only grow. Damage growth will create a growth of the fully damaged zone W_c (and thus a negative velocity v_n). The last term in (62) is thus automatically positive. Whether this term is zero or not depends on the regularity of ψ on the boundary G_c. This regularity must be assessed from the non-local constitutive model condition: $\bar{y} - Y_c \leq 0$.

Note that dissipation may also be written

$$D = \int_{\Omega^-} Y \dot{d} \, d\omega + \int_{\Omega^+} \overline{Y} \dot{\overline{d}} \, d\omega - \int_{\Gamma_c} \psi \, v_n \, da$$

(63)

where $\overline{y}$ is defined by (25) (y replaced by Y). The above expression exhibits the duality between $\overline{y}$ and $\dot{\overline{d}}$ in the localization zone.

RESULTS

We consider a 1D axisymmetric fiber pull-out depicted in Figure 4. The fiber of radius r_i is considered rigid and infinitely long. It is pulled out of a clamped circular domain of radius $r_e = r_i + L$. The only non-zero stress component is the shear stress τ satisfying the following equilibrium conditions

$$(\tau r)_{,r} = 0 \Rightarrow \tau(r) = \frac{\tau(r_i)r_i}{r}$$

(64)

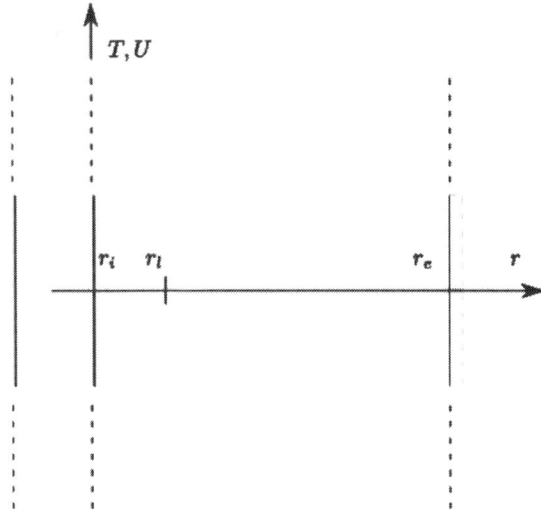

Figure 4: Pull-out of an infinite fiber of radius r_i from a tube of radius r_e. Radius r_l indicates the (evolving) extent of the non-local damage zone.

The only non-zero strain is the shear strain, derivative of the displacement along the fiber direction

$$\gamma = u,_r$$

(65)

We consider the following free energy density involving some hardening function $h(d)$, satisfying $h(1)=0$. The shear stiffness is denoted μ and Y_c is also a material parameter.

$$\psi(\gamma,d) = \frac{1}{2}\mu(1-d)\gamma^2 + Y_c h(d)$$

(66)

So, state laws read

$$\tau = \mu(1-d)\gamma, \quad Y = \frac{1}{2}\mu\gamma^2 - Y_c h'$$

(67)

The local evolution model is given by

$$\dot{d} \geq 0, \quad Y - Y_c \leq 0, \quad (Y - Y_c)\dot{d} = 0$$

(68)

The condition $Y=Y_c$ reduces to

$$\frac{\tau}{\tau_c} = \underbrace{(1-d)\sqrt{h'+1}}_{g(d)}, \quad \tau_c = \sqrt{2\mu Y_c}$$

(69)

Let us now be more precise on the type of function $g(d)$ we will be considering. Basically, we are interested by C^1 positive concave functions with a maximum value at some damage $d_c < 1$:

$$g(d) \in C^1([0,1]) : g'' < 0, \ g(0) = 1, \ g(1) = 0, \ g'(d_c) = 0$$

(70)

shall use in what follows

$$g(d) = (1-d)\exp\left(\frac{d}{1-d_c}\right)$$

(71)

The corresponding stress strain curve is given in Figure 5. We will now search for the complete solution linking the (non-dimensional) shear stress T needed to move by a (non-dimensional) displacement U the fiber:

$$T = \frac{\tau(r_i)}{\tau_c}, \quad U = \frac{u(r_i)\mu}{r_i\tau_c} \tag{72}$$

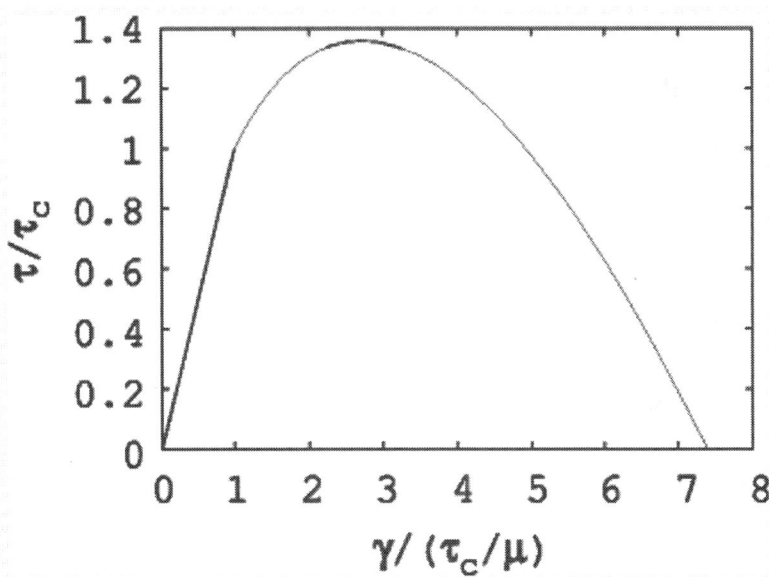

Figure 5: Local constitutive model: stress versus rising strain (case d_c=0.5).

Four regimes will be observed. They are depicted in Figure 6: elastic, local damage, local and non-local damage and finally purely non-local damage. The first two regimes may be solved analytically whereas the two last one may not. We however pursue as much as possible the analytical path. Next section is devoted to a numerical solver.

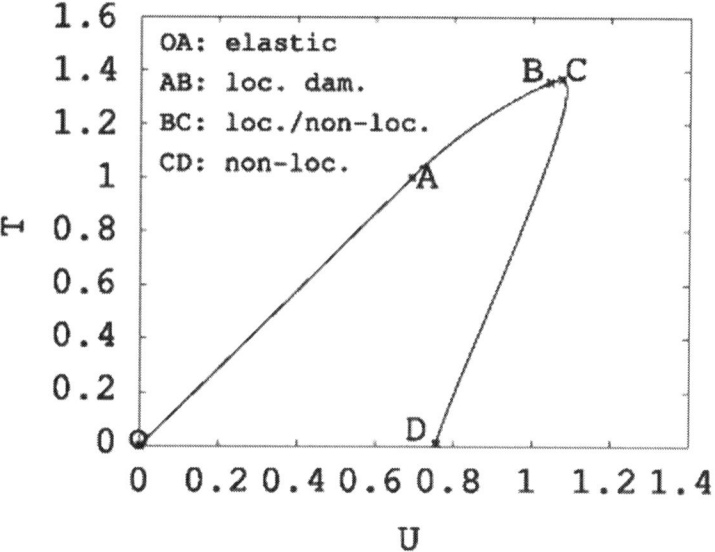

Figure 6: Force-displacement curve in the case n = 2. The elastic part of the solution is on part [OA] of the curve. The part [AB] corresponds to the development of local damage whereas part [BC] corresponds to both local and non-local damage development. Finally, part [CD] is governed only by non-local damage. Letters A, B and C are located at $T_A=1, T_B=1.351$ and $T_C=1.368$, respectively. The latter being the limit load.

Pure Elastic Regime: $T \in [0,T_A=1]$

The displacement solution is given by

$$u(r) = T\frac{\tau_c}{\mu} \int_r^{r_e} \frac{r_i}{r}\, dr = T\frac{\tau_c}{\mu} r_i \log\left(\frac{r_e}{r}\right)$$

(73)

We thus have a linear relationship between the stress and displacement

$$T = U\left(\log\left(\frac{r_e}{r_i}\right)\right)^{-1}$$

(74)

Local Damage Regime: $T \in [1, T_B]$

When T reaches 1 local damage starts around the fiber. Its distribution is obtained by combining (64) and (69)

$$T = g(d)\frac{r}{r_i}$$

(75)

This distribution of local damage is acceptable provided the condition below holds true

$$\|\nabla\phi\| \leq 1 \text{ i.e. } \|\nabla d\| \leq d'(\phi(d))$$

(76)

The norm of the damage gradient is maximum at $r=r_i$ and of value

$$\left|\frac{dd}{dr}\right| = \frac{1}{r_i}\frac{g(d_i)}{g'(d_i)}$$

(77)

where $d_i=d(r_i)$. Considering a general power law damage profile (13), the condition (76) is

$$\left|\frac{g(d_i)}{g'(d_i)}\right| \leq \frac{r_i}{l_c}n(1-d_i)^{1-1/n}$$

(78)

Let us denote by d_i^B the smallest value of damage for which the condition above is violated and T_B the corresponding loading. Due to the fact that $g'(d_c)=0$, it is clear that d_i^B will be slightly lower than d_c. We note that as the material length gets bigger with respect to r_i, non-locality (violation of (78)) will step in for smaller and smaller damage d_i. Considering the choice (71), we get the condition

$$\left|\frac{(1-d_i)(1-d_c)}{d_c-d_i}\right| \leq \frac{r_i}{l_c}n(1-d_i)^{1-1/n}$$

(79)

For instance if $n=1$, we get

$$d_i^B = \frac{d_c - (1 - d_c)(l_c/r_i)}{1 - (1 - d_c)(l_c/r_i)}$$

(80)

As a numerical application, with d_c=0.5 and l_c/r_i=0.1, we get d_i^B =0.47.

Combined Local and Non-local Damage Regime: $T \in [T_B, T_C]$

For a loading higher than T_B, non-local damage will develop close to the fiber. Let $[r_i, r_l]$ be the current extension of the non-local damage zone in which damage ranges from d_i to d_l following:

$$r(d, d_i) = r_i + \phi(d_i) - \phi(d)$$

(81)

The condition for the non-local zone to grow is $\overline{Y} - Y_c$, i.e.

$$\int_{r_i}^{r_l} (Y - Y_c) \frac{dd}{d\phi} r \, dr = 0$$

(82)

Using, (81), we may rewrite it as

$$\int_{d_l}^{d_i} (Y - Y_c) r(d, d_i) \, dd = 0$$

(83)

Now, using (69), we get

$$T = \left(\frac{\int_{d_l}^{d_i} (1 - d)^{-2} g^2(d)(r(d, d_i)/r_i) \, dd}{\int_{d_l}^{d_i} (1 - d)^{-2} (r(d, d_i)/r_i)^{-1} \, dd} \right)^{1/2}$$

(84)

Since loading is rising, so does local d_l damage at $r=r_l$, following

$$T = \frac{r_l}{r_i} g(d_l)$$

(85)

Given T, system (84)-(85) returns unknowns di and dl as well as the extent of the non-local zone $r_l=r(d_l,d_i)$. We note that for $T=T_B$, we have

$d_j = d_i$ and $r_j = r_i$. Let T_C be the loading above which the system has no solution.

Non-local Damage Regime: T Decreases from T_C to 0

There is no solution of the problem for a loading higher than T_C. When the loading decreases below T_C, there is of course a possible elastic solution. Another possible solution is the further development of the non-local damage zone (while damage in the local zone no longer evolves since loading is decreasing). The system of equations to solve still involves (84)

$$T = \left(\frac{\int_{d_l}^{d_i} (1-d)^{-2} g^2(d)(r(d,d_i)/r_i)\, dd}{\int_{d_l}^{d_i} (1-d)^{-2} (r(d,d_i)/r_i)^{-1}\, dd} \right)^{1/2} \tag{86}$$

Equation (85) is now different and reads

$$T_C = \frac{r_l}{r_i} f(d_l) \tag{87}$$

Indeed the damage at r_l did not change from its value at load T_C because the load has been decreasing afterwards.

Analysis of the Displacement of the Fiber

The displacement of the fiber is given by

$$U = T \int_{r_i}^{r_e} \frac{1}{(1-d)r}\, dr \tag{88}$$

As the damage around the fiber goes to 1, the integrand goes to infinity. But, at the same time the loading goes to zero. Let us study the limit of the fiber displacement for the loading going to zero. The loading is given by (84) recalled below

$$T = \left(\frac{\int_{d_i}^{d_i} (1-d)^{-2} g^2(d)(r(d,d_i)/r_i)\, dd}{\int_{d_i}^{d_i} (1-d)^{-2} (r(d,d_i)/r_i)^{-1}\, dd} \right)^{1/2} = \left(\frac{\int_{d_i}^{d_i} N(d,d_i)\, dd}{\int_{d_i}^{d_i} D(d,d_i)\, dd} \right)^{1/2} \tag{89}$$

Due to the property of $g(d)$, (70), we have

$$g(d) = O(1 - d) \text{ as } d \to 1$$

(90)

So

$$0 < N(d, d_i) < +\infty, \forall d, d_i \in [0, 1]$$

(91)

Finally, we have

$$T = O\left(\sqrt{1 - d_i}\right) \text{ as } d_i \to 1$$

(92)

Note that this property does not depend on the choice of $d(\phi)$. Going back to the displacement expression, (88), we have

$$U = T \int_{d_l}^{d_i} (1 - d)^{-1} (r(d, d_i))^{-1} \frac{dr(d, d_i)}{dd} \, dd + CT$$

(93)

where C is a finite constant and

$$r(d, d_i) = r_i + \phi(d_i) - \phi(d) = r_i + l_c \left((1 - d)^{1/n} - (1 - d_i)^{1/n}\right)$$

(94)

assuming a power law asymptotic behavior of $\phi(d)$ as d goes to 1:

$$\phi(d) = l_c(1 - (1 - d)^{1/n})$$

(95)

Finally, we get

$$U = T\, O\left((1 - d_i)^{1/n - 1} + C\right) = O((1 - d_i)^{1/n - 1/2}) \text{ as } d_i \to 1$$

(96)

We conclude that there exists three regimes of delocalization.

$$\lim_{d_i \to 1} U = 0 \qquad \text{if } n < 2$$

(97)

$$0 < \lim_{d_i \to 1} U < +\infty \quad \text{if } n = 2$$

(98)

$$\lim_{d_i \to 1} U = +\infty \quad \text{if } n > 2$$

(99)

When $n<2$, the fiber displacement must be zero for total failure.

When $n=2$, there exists a limit value of fiber displacement before total failure and when $n>2$, it takes an infinite displacement before total failure. It is interesting to note that these three regimes also exist in gradient damage models [20].

Numerical Solve

Last section gave some insight on the different regimes in the solution. In order to plot the solution, we detail here a 1D numerical solver. This code is rather ad hoc for 1D problem, since we force the advance of the Γ boundary and find the corresponding loading and fields. General 2D and 3D solvers will be detailed in a forthcoming paper. We search for the solution at a set of discrete times. Consider the solution known at time t_n, the solution at time t_{n+1} must satisfy the following equations.

Kinematics and Equilibrium on $]r_i, r_e[$

$$u^{n+1} \quad \in \quad \mathcal{U} = \left\{ u \in H^1(] r_i, r_e[) : u(r_e) = 0 \right\} \tag{100}$$

$$\gamma\left(u^{n+1}\right) \quad = \quad u^{n+1}_{,r} \tag{101}$$

$$\int_{r_i}^{r_e} \tau^{n+1} \, \gamma(u^*) \, r \, dr = \tau_c T^{n+1} u^*(r_i) r_i, \quad \forall u^* \in \mathcal{U} \tag{102}$$

State Laws and d (ϕ) Relation on $]r_i, r_e[$

$$\tau^{n+1} = \tau\left(\gamma\left(u^{n+1}\right), d^{n+1}\right) = \left(1 - d^{n+1}\right) \mu \gamma\left(u^{n+1}\right) \tag{103}$$

$$Y^{n+1} = Y\left(\gamma\left(u^{n+1}\right), d^{n+1}\right) = \frac{1}{2} \mu \gamma\left(u^{n+1}\right)^2 - Y_c h'\left(d^{n+1}\right) \tag{104}$$

$$d^{n+1} = d\left(\phi^{n+1}\right) \tag{105}$$

Non-local Evolution Law on $]\, r_i, r_l^{n+1}\, [$

$$a = \int_{r_i}^{r_l^{n+1}} \left(Y^{n+1} - Y_c\right) d'\left(\phi^{n+1}\right) r\, dr \leq 0,$$

(106)

$$b = \phi^{n+1}(r_i) - \phi^n(r_i) \geq 0, \quad ab = 0$$

(107)

$$\phi^{n+1} = \phi^{n+1}\left(r_l^{n+1}\right) + r_l^{n+1} - r, \quad \text{on }]\, r_i, r_l^{n+1}\, [$$

(108)

Local Evolution Law on $]\, r_l^{n+1}, r_e\, [$

$$Y^{n+1} - Y_c \leq 0, \quad \phi^{n+1} - \phi^n \geq 0, \quad \left(Y^{n+1} - Y_c\right)\left(\phi^{n+1} - \phi^n\right) = 0$$

(109)

Regarding space discretization, the segment $]r_i, r_e[$ is discretized with a set of finite elements. Initially, the non-local zone is empty and we proceed with a classical Newton-Raphson scheme depicted in the solver flowchart without non-local zone.

Solver Flowchart without Non-local Zone

- initialization: $u^0 = d^0 = T^0 = 0$
- elastic step: find the load step for which damage starts
- load step $T^{n+1} = T^n + DT$
- iterations initialization $k=0$
- solve linear system (110) to find Du
- after the first iteration adapt the load step so that the maximum damage increment is d_{inc}
- update using (111)-(118)
- if residual $\leq$ tol, go to 9 else go to 5

- if $\left\|\nabla\phi^{n+1}\right\| \leq 1$, go to 3, else go to solver flowchart with non-local zone

The linear problem at each iteration reads: find $\Delta u \in U$ such that:

$$\int_{r_i}^{r_e} H^k \gamma(\Delta u) \gamma(u^*) r \, dr = \tau_c T^{n+1} u^*(r_i) r_i - \int_{r_i}^{r_e} \tau^k \gamma(u^*) r \, dr, \forall u^* \in \mathcal{U}$$

(110)

where the right hand side is the residual at iteration k. Once the displacement correction is obtained, the local update of the fields is computed from

$$u^{k+1} = u^k + \Delta u$$

(111)

$$\tau^{k+1} = \tau\left(\gamma\left(u^{k+1}\right), d^{k+1}\right), \quad Y^{k+1} = Y\left(\gamma\left(u^{k+1}\right), d^{k+1}\right)$$

(112)

$$Y^{k+1} - Y_c \le 0, d^{k+1} - d^n \ge 0, \left(Y^k - Y_c\right)\left(d^k - d^n\right) = 0$$

(113)

$$\phi^{k+1} = \phi\left(d^{k+1}\right)$$

(114)

whereas tangent operators are obtained by

$$H^{k+1} = H_{\gamma\gamma}^{k+1} - \eta^k H_{\gamma d}^{k+1} \left(H_{dd}^{k+1}\right)^{-1} H_{d\gamma}^{k+1}$$

(115)

$$\eta^{k+1} = 1, \text{ if } d^{k+1} - d^k > 0 \text{ and } 0 \text{ otherwise}$$

(116)

$$H_{\gamma\gamma}^{k+1} = \frac{\partial \tau}{\partial \gamma} |_{k+1}, \quad H_{\gamma d}^{k+1} = \frac{\partial \tau}{\partial d} |_{k+1},$$

(117)

$$H_{d\gamma}^{k+1} = -\frac{\partial Y}{\partial \gamma} |_{k+1}, H_{dd}^{k+1} = -\frac{\partial Y}{\partial d} |_{k+1}$$

(118)

At the end of each load step, the gradient of the level set is computed. If it is below 1 everywhere the next load step is applied. If not, a non-local zone is placed and the solver flowchart with non-local zone is used.

Solver Flowchart with Non-local Zone

- initialization: $r_i^0 = r_i$
- increase non-local zone: $r_i^{n+1} = r_i^n + \Delta r_i$
- iterations initialization: $k=0$
- linear solve: solve (119) to find Du, DT, Df
- load update: $T^{k+1} = T^k + DT$
- update in local zone (111)-(118), and non-local zone (120)-(124)
- if residual $\leq$ tol, go to 8, else go to 4
- if domain not fully broken ($d(r_i) < 1$), go to 2, else go to 9
- end

The extent of the non-local zone is imposed and one tries to find a continuous displacement and damage field satisfying the problem.

The linear symmetric problem to be solved at each iteration when the non-local zone is not empty is to find $\Delta u \in U, \Delta \phi \in A \ \ \Delta T \in R$ such that

$$\int_{r_i}^{r_i^{n+1}} \left(H_{\gamma\gamma}^k \gamma(\Delta u) + H_{\gamma d}^k d'\left(\phi^k\right) \Delta \phi \right) \gamma(u^*) r \, dr + \int_{r_i^{n+1}}^{r_e} H^k \gamma(\Delta u) \gamma(u^*) r \, dr - \tau_c \Delta T u^*(r_i) =$$

$$\tau_c T^k u^*(r_i) r_i - \int_{r_i}^{r_e} \tau^k \gamma(u^*) r \, dr, \forall u^* \in U$$

$$\int_{r_i}^{r_i^{n+1}} \left(H_{d\gamma}^k d'\left(\phi^k\right) \gamma(\Delta u) + \left(H_{dd}^k d'^2\left(\phi^k\right) + \left(Y_c - Y^k\right) d''\left(\phi^k\right) \right) \Delta \phi \right) \phi^* r \, dr =$$

$$\int_{r_i}^{r_i^{n+1}} \left(Y^k - Y_c\right) d'\left(\phi^k\right) \phi^* r \, dr, \quad \forall \phi^* \in A$$

$$-\Delta \phi + \eta^k \frac{\left(H_{dd}^k\right)^{-1} H_{d\gamma}^k \gamma(\Delta u)}{d'(\phi^k) \mid_{r_i^{n+1,+}}} = \left(\phi^k - \phi\left(d^k\right)\right) \mid_{r_i^{n+1,+}}$$

$$(119)$$

Where h^k is evaluated following (116). The update in the local zone follows (111)-(118) whereas in the non-local zone we have

$$u^{k+1} = u^k + \Delta u, \quad \phi^{k+1} = \phi^k + \Delta \phi, \tag{120}$$

$$d^{k+1} = d\left(\phi^{k+1}\right) \tag{121}$$

$$\tau^{k+1} = \tau\left(\gamma\left(u^{k+1}\right), d^{k+1}\right), \quad Y^{k+1} = Y\left(\gamma\left(u^{k+1}\right), d^{k+1}\right) \tag{122}$$

$$H_{\gamma\gamma}^{k+1} = \frac{\partial \tau}{\partial \gamma}\bigg|_{k+1}, \quad H_{\gamma d}^{k+1} = \frac{\partial \tau}{\partial d}\bigg|_{k+1},$$

(123)

$$H_{d\gamma}^{k+1} = -\frac{\partial Y}{\partial \gamma}\bigg|_{k+1}, H_{dd}^{k+1} = -\frac{\partial Y}{\partial d}\bigg|_{k+1}$$

(124)

Is is interesting to note the difference between the two solver flowcharts. When the non-local zone is empty, the linear solve deals only with displacement increments and the local update deals with the damage variable. On the contrary, when the non-local zone is not empty, the linear solve involves both displacement and damage (or more precisely the surrogate ϕ variable) increment in the non-local zone (local zone being treated as before).

The mesh is built so that it is much finer in the localization zone. Node j is located at a position $x(j)$ given by

$$x(j) = \frac{(r_i + ((j-1)/N)(r_e - r_i))^2}{r_e - r_i} + r_i, \quad j = 1, \ldots, N+1$$

(125)

where N is the number of elements considered. Results will be shown for the following mechanical parameters:

$$r_i = 0.1m, \quad r_e = 0.2m, \quad l_c = 0.02m, \quad \frac{\tau_c}{\mu} = 10^{-4}, \quad d_c = 0.5$$

(126)

and numerical parameters

$$N = 200, \quad d_{\text{inc}} = 0.02$$

(127)

Regarding parameter Dr_l, the non-local zone is advanced by one element at a time or smaller when damage gets close to 1 at r_i. This is done in order to capture the full load-displacement curve. The formula used in the simulation is

$$\Delta r_l = \max\left(\min\left(h^*, (l_c - \phi^n(r_i))/2\right), 1.e^{-12} * l_c\right)$$

(128)

where h^* is the size of the element adjacent to the non-local zone at time step n. The initial ($n=0$) non-local zone needs to be more than one-element for convergence. Between 5 and 10 elements are used.

As a final remark on the solver flowchart with the non-local zone, we noticed that in non-local zone update, it was more efficient (reduced number of iterations) to take $D\phi$ as the one ensuring damage continuity rather that picking the one coming from step 3.

DISCUSSION

In Figure 6, the force-displacement curve in the case $n = 2$ is shown. The figure indicates the different regime of the solution (pure elastic, local damage, coupled and pure non-local damage). Note that snap-back is taken into account automatically since the loading is not imposed but an unknown in the numerical scheme. Profiles of ϕ and damage along the radius at different loads are depicted in Figure 7.

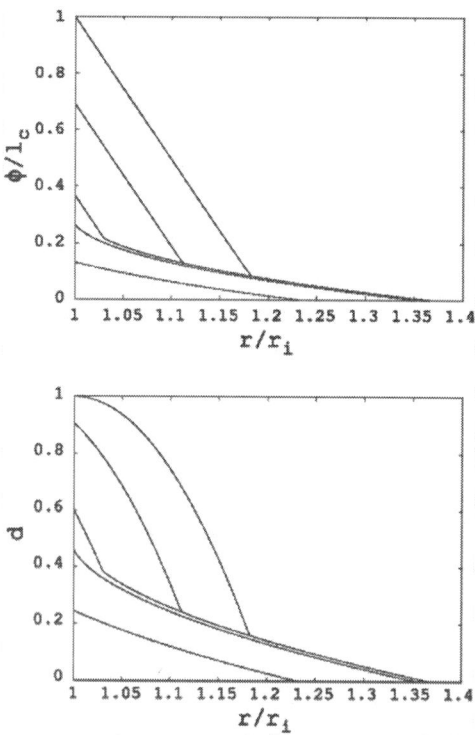

Figure 7: Distributions of ϕ/l_c (top figure) and damage (bottom figure) for different loadings. Details on the curves in each plot from bottom

to top: Bottom curve corresponds to $T=1.232$, damage evolution is purely local. The next curve is for $T=1.351$. It corresponds to the load at which $\|\nabla\phi\|=1$ at $r=r_i$ and non-locality steps in. Next curve is for $T=1.368$, it is the limit load. The load then decreases and damage evolution is purely non-local. Last two curves are for $T=1.012$ and $T=0.02$, respectively. The latter case depicts the profile at complete decohesion of the fiber.

Figure 8 shows the influence of the delocalization parameter n. Plots confirm the analytical limit results (97). As long as damage is purely local, all curves are superposed. As non-locality steps in, the delocalization parameters n plays a role.

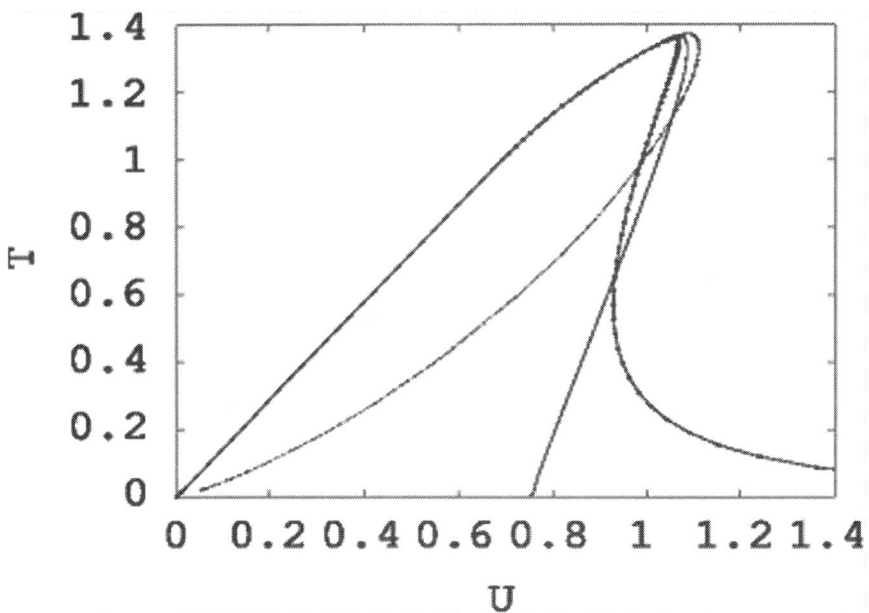

Figure 8: Force-displacement curves for $n = 1$ (small dots), $n = 2$ (solid line), $n = 3$ (big dots).

Finally, in order to show the insensitivity of the model with respect to the discretization parameters N, d_{inc}, we show Figure 9 the influence of the choice of the N parameter (for the case $n=2$ and $d_{inc}=0.02$). In Figure 10, we show the influence of d_{inc} (for the case $n=2$ and $N=50$). Note that as expected, parameter d_{inc} has only an influence when damage is purely local (rising part of the curve). For both figures, a zoom was used. Otherwise, curves cannot be distinguished.

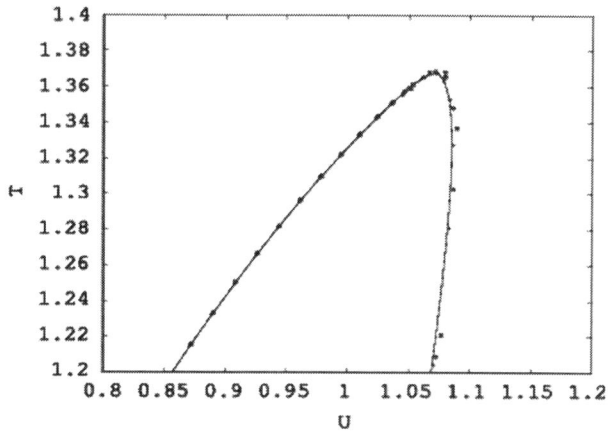

Figure 9: Force-displacement curves (zoom) for n = 2 and d_{inc}=0.02 with different mesh sizes: N = 200 (solid line), N = 20 stars, N = 30 circles, N = 40 plus sign.

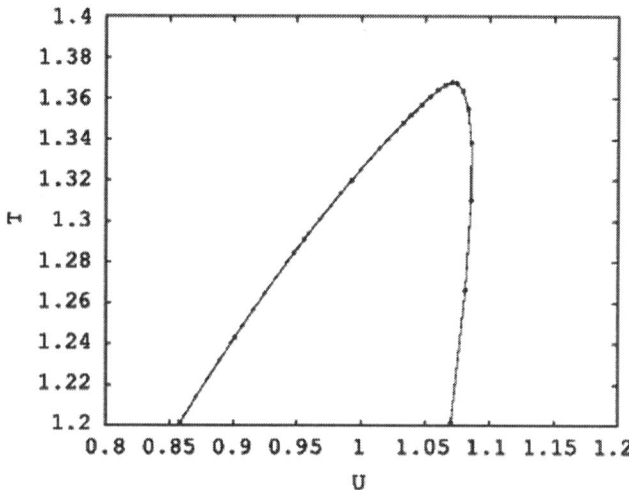

Figure 10: Force-displacement curves (zoom) for n = 2 and N = 50 with different values of d_{inc}:d_{inc}=0.02 (solid line),d_{inc}=0.1 stars,d_{inc}=0.05 circles,d_{inc}=0.01 plus sign.

CONCLUSIONS

The Thick Level Set damage model allows coupling local damage evolution in some part of the domain to a non-local damage evolution in the localization zone. Damage gradient is bounded. The bound is reached in the non-local zone (localization zone) and not reached in the local one. The localization zone boundary is the main unknown in the model. It evolves ensuring damage continuity. A semi-analytical 1D solution has been developed showing different regimes in the solution (elastic, local damage, coupled local and non-local damage and finally pure non-local damage). The solution was plotted using a numerical scheme. This numerical scheme is ad hoc for the 1D problem considered. The corresponding numerical implementation for 2D and 3D cases will be the subject of a forthcoming publication.

AUTHORS CONTRIBUTIONS

All authors contributed to the main ideas in the paper: the way to couple local and non-local evolutions of damage. NM came up with the analytical solution. NM and NC designed the 1D code to plot the results. All authors read and approved the final manuscript.

ACKNOWLEDGEMENTS

The support of the ERC Advanced Grant XLS no 291102 is greatfully acknowledged. Professor Antonio Huerta is also acknowledged for his advice.

REFERENCES

1. Moës N, Stolz C, Bernard P-E, Chevaugeon N (2011) A level set based model for damage growth: the thick level set approach. Int J Numer Meth Eng 86:358-380

2. Bernard P-E, Moës N, Chevaugeon N (2012) Damage growth modeling using the Thick Level Set (TLS) approach: efficient

discretization for quasi-static loadings. Comput Meth Appl Mech Eng 233¿236:11-27

3. Stolz C, Moës N (2012) A new model of damage: a moving thick layer approach. Int J Fract 174:49-60

4. Stolz C, Moës N (2012) On the rate boundary value problem for damage modelization by Thick Level Set. In: ACOME 2012 Proceeding. Ho-Chi-Minh, Viet Nam. [http://hal.archives-ouvertes.fr/hal-00725635] http://hal.archives-ouvertes.fr/hal-00725635

5. Karma A, Kessler D, Levine H (2001) Phase-field model of mode III dynamic fracture. Phys Rev Lett 87(4):045501

6. Miehe C, Welschinger F, Hofacker M (2010) Thermodynamically consistent phase-field models of fracture: Variational principles and multi-field FE implementations. Int J Numer Meth Eng 83(10):1273-1311

7. Spatschek R, Brener E, Karma A (2011) Phase field modeling of crack propagation. Phil Mag 91(1):75-95

8. Francfort GA, Marigo J-J (1998) Revisiting brittle fracture as an energy minimization problem. J Mech Phys Solid 46:1319-1412

9. [http://link.springer.com/10.1007/s10659-007-9107-3] Bourdin B, Francfort GA, Marigo J-J (2008) The Variational Approach to Fracture, 5¿148,

10. Comi C, Mariani S, Perego U (2007) An extended FE strategy for transition from continuum damage to mode I cohesive crack propagation. Int. J. Numer. Anal. Meth. Geomech 31:213-238

11. Sethian JA (1999) Level set methods and fast marching methods: evolving interfaces in computational geometry, fluid mechanics, computer vision and material science. Cambridge University Press, UK.

12. Maugin GA (1990) Internal variables and dissipative structures. J Non-Equilibrium Therm 15:173-192

13. Frémond M, Nedjar B (1996) Damage, gradient of damage and principle of virtual power. Int J Solid Struct 33(8):1083-1103

14. Comi C (1999) Computational modelling of gradient-enhanced damage in quasi-brittle materials. Mech Cohesive-Frictional Mater 36(April 1997):17-36

15. Pijaudier-Cabot G, Bazant ZP (1987) Nonlocal dalmage theory. J Eng Mech ASCE 113:1512-1533

16. Bazant ZP, Jirasek M (2002) Nonlocal integral formulations of plasticity and damage: survey of progress. J Eng Mech 128(November):1119-1149

17. Lubineau G, Azdoud Y, Han F, Rey C, Askari A (2012) A morphing strategy to couple non-local to local continuum mechanics. J Mech Phys Solid 60(6):1088-1102

18. Azdoud Y, Han F, Lubineau G (2013) A Morphing framework to couple non-local and local anisotropic continua. Int J Solid Struct 50(9):1332-1341

19. Lions P-L (1982) Generalized solutions of Hamilton-Jacobi equations. Pitman Advanced Publishing Program, Boston.

20. Lorentz E, Godard V (2011) Gradient damage models: toward full-scale computations. Comput Meth Appl Mech Eng 200(21¿22):1927-1944

21. Chung-Min L, Rubinstein J (2006) Elliptic equations with diffusion coefficient vanishing at the boundary: theoretical and computational aspects. Quaterly Appl Math 64:725-747

22. Pradeilles-Duval RM, Stolz C (1995) Mechanical transformations and discontinuities along a moving surface. J Mech Phys Solid 43(1):91-121

23. Hill R (1986) Energy-momentum tensors in elastostatics: some reflections on the general theory. J. Mech. Phys. Solids 34(3):305-317

24. Stolz C (2010) On micro-macro transition in non-linear mechanics. Materials 3(1):296-317

25. Rockafellar RT (1970) Convex analysis. Princeton University Press, USA.

Citations

CHAPTER 1

Ali Abdul-Aziz, Frank Abdi, Ramakrishna T. Bhatt, and Joseph E. Grady, "Durability Modeling of Environmental Barrier Coating (EBC) Using Finite Element Based Progressive Failure Analysis," Journal of Ceramics, vol. 2014, Article ID 874034, 10 pages, 2014. doi:10.1155/2014/874034.

CHAPTER 2

Uzair Ahmed Dar, Weihong Zhang, and Yingjie Xu, "FE Analysis of Dynamic Response of Aircraft Windshield against Bird Impact," International Journal of Aerospace Engineering, vol. 2013, Article ID 171768, 12 pages, 2013. doi:10.1155/2013/171768.

CHAPTER 3

M. Alemi-Ardakani, A. S. Milani, S. Yannacopoulos, et al., "Microtomographic Analysis of Impact Damage in FRP Composite Laminates: A Comparative Study," Advances in Materials Science and Engineering, vol. 2013, Article ID 521860, 10 pages, 2013. doi:10.1155/2013/521860.

CHAPTER 4

Alexander Maier, Roland Schmidt, Beate Oswald-Tranta, and Ralf Schledjewski, Non-Destructive Thermography Analysis of Impact Damage on Large-Scale CFRP Automotive Parts, doi:10.3390/ma7010413.

CHAPTER 5

Irina Severin, Rochdi El Abdi, Guillaume Corvec, and Mihai Caramihai, Optical Fiber Embedded in Epoxy Glass Unidirectional Fiber Composite System, doi:10.3390/ma7010044.

CHAPTER 6

Yang Yang, Juntao Chen, and Ming Xiao, "Analysis of Seismic Damage of Underground Powerhouse Structure of Hydropower Plants Based on Dynamic Contact Force Method," Shock and Vibration, vol. 2014, Article ID 859648, 13 pages, 2014, doi:10.1155/2014/859648.

CHAPTER 7

Gautam S. Chandekar, Bhushan S. Thatte, and Ajit D. Kelkar, "On the Behavior of Fiberglass Epoxy Composites under Low Velocity Impact Loading," Advances in Mechanical Engineering, vol. 2010, Article ID 621406, 11 pages, 2010. doi:10.1155/2010/621406.

CHAPTER 8

Guoqing Chen, Yan Zhang, Runqiu Huang, Fan Guo, and Guofeng Zhang, "Failure Mechanism of Rock Bridge Based on Acoustic Emission Technique," Journal of Sensors, Article ID 964730, in press.

CHAPTER 9

Nicolas Moës, Claude Stolz, and Nicolas Chevaugeon, Coupling Local and Non-local Damage Evolutions with the Thick Level Set model, doi:10.1186/s40323-014-0016-2.

Index